HSC Year 12
MATHEMATICS
ADVANCED

SARAH HAMPER

SERIES EDITOR: ROBERT YEN

2020 UPDATED SYLLABUS · 2020 UPDATED SYLLABUS · 2020 UPDATED SYLLABUS · 2020 UPDATED SYLLABUS

+ topic summaries
+ graded practice questions
  with worked solutions
+ HSC exam topic grids (2011–2020)

STUDY
NOTES

A+ HSC Mathematics Advanced Study Notes
1st Edition
Sarah Hamper
ISBN 9780170459228

Publishers: Robert Yen, Kirstie Irwin
Project editor: Tanya Smith
Cover design: Nikita Bansal
Text design: Alba Design
Project designer: Nikita Bansal
Permissions researcher: Corrina Gilbert
Production controller: Karen Young
Typeset by: Nikki M Group Pty Ltd

Any URLs contained in this publication were checked for currency during the production process. Note, however, that the publisher cannot vouch for the ongoing currency of URLs.

NSW Education Standards Authority (NESA) Higher School Certificate Examination Mathematics Advanced: 2019, 2020; NSW Education Standards Authority (NESA) 2017 Higher School Certificate Examination Mathematics; NSW Education Standards Authority (NESA) 2015 Higher School Certificate Examination Mathematics General 2; NSW Education Standards Authority (NESA) 2013 Higher School Certificate Examination General Mathematics © NSW Education Standards Authority for and on behalf of the Crown in right of the State of New South Wales.

For product information and technology assistance,
in Australia call **1300 790 853**;
in New Zealand call **0800 449 725**

For permission to use material from this text or product, please email **aust.permissions@cengage.com**

ISBN 978 0 17 045922 8

Cengage Learning Australia
Level 7, 80 Dorcas Street
South Melbourne, Victoria Australia 3205

Cengage Learning New Zealand
Unit 4B Rosedale Office Park
331 Rosedale Road, Albany, North Shore 0632, NZ

For learning solutions, visit **cengage.com.au**

Printed in China by 1010 Printing International Limited.
1 2 3 4 5 6 7 25 24 23 22 21

ABOUT THIS BOOK

Introducing *A+ HSC Year 12 Mathematics*, a new series of study guides designed to help students revise the topics of the new HSC maths courses and achieve success in their exams. *A+* is published by Cengage, the educational publisher of *Maths in Focus* and *New Century Maths*.

For each HSC maths course, Cengage has developed a STUDY NOTES book and a PRACTICE EXAMS book. These study guides have been written by experienced teachers who have taught the new courses, some of whom are involved in HSC exam marking and writing. This is the first study guide series to be published after students sat the first HSC exams of the new courses in 2020, so it incorporates the latest changes to the syllabus and exam format.

This book, *A+ HSC Year 12 Mathematics Advanced Study Notes,* contains topic summaries and graded practice questions, grouped into 7 broad topics, addressing the outcomes in the Mathematics Advanced syllabus. The topic-based structure means that this book can be used for revision after a topic has been covered in the classroom, as well as for course review and preparation for the trial and HSC exams. Each topic chapter includes a review of the main mathematical concepts and multiple-choice and short-answer questions with worked solutions. Past HSC exam questions have been included to provide students with the opportunity to see how they will be expected to show their mathematical understanding in the exams. An HSC exam topic grid (2011–2020) guides students to where and how each topic has been tested in past HSC exams.

Mathematics Advanced Year 12 topics

1. Graphing functions
2. Trigonometric functions
3. Differentiation
4. Integration
5. Series, investments, loans and annuities
6. Statistics and bivariate data
7. Probability distributions

This book contains:

- Concept map (see p. 2 for an example)
- Glossary and digital flashcards (see p. 3 for an example)
- Topic summary, addressing key outcomes of the syllabus (see p. 4 for an example)
- Practice set 1: 20 multiple-choice questions (see p. 9 for an example)
- Practice set 2: 20 short-answer questions (see p. 16 for an example)
- Questions graded by level of difficulty: foundation ▢▢▢, moderate ▢▢▢, complex ▢▢▢
- Worked solutions to both practice sets (see p. 20 for an example)
- HSC exam topic grid (2011–2020) (see p. 25 for an example)

The companion A+ PRACTICE EXAMS book contains topic exams and practice HSC exam papers, both of which are written and formatted in the style of the HSC exam, with spaces for students to write answers. Worked solutions are provided, along with the author's expert comments and advice, including how each exam question is marked. As a special bonus, the worked solutions to the 2020 HSC exam paper have been included.

This A+ STUDY NOTES book will become a staple resource in your study in the lead-up to your final HSC exam. Revisit it throughout Year 12 to ensure that you do not forget key concepts and skills. Good luck!

CONTENTS

CHAPTER 1

GRAPHING FUNCTIONS

CHAPTER 2

TRIGONOMETRIC FUNCTIONS

CHAPTER 3

DIFFERENTIATION

9780170459228

YEAR 12 COURSE OVERVIEW

See each concept map printed in full size at the beginning of each chapter.

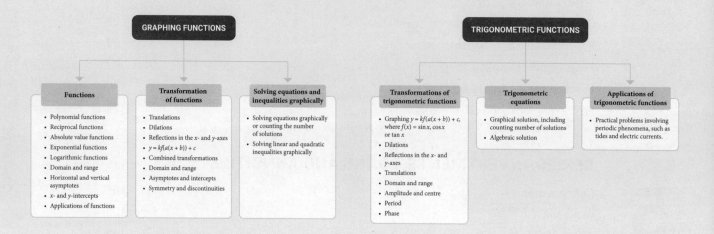

GRAPHING FUNCTIONS

Functions
- Polynomial functions
- Reciprocal functions
- Absolute value functions
- Exponential functions
- Logarithmic functions
- Domain and range
- Horizontal and vertical asymptotes
- x- and y-intercepts
- Applications of functions

Transformation of functions
- Translations
- Dilations
- Reflections in the x- and y-axes
- $y = kf(a(x + b)) + c$
- Combined transformations
- Domain and range
- Asymptotes and intercepts
- Symmetry and discontinuities

Solving equations and inequalities graphically
- Solving equations graphically or counting the number of solutions
- Solving linear and quadratic inequalities graphically

TRIGONOMETRIC FUNCTIONS

Transformations of trigonometric functions
- Graphing $y = kf(a(x + b)) + c$, where $f(x) = \sin x, \cos x$ or $\tan x$
- Dilations
- Reflections in the x- and y-axes
- Translations
- Domain and range
- Amplitude and centre
- Period
- Phase

Trigonometric equations
- Graphical solution, including counting number of solutions
- Algebraic solution

Applications of trigonometric functions
- Practical problems involving periodic phenomena, such as tides and electric currents.

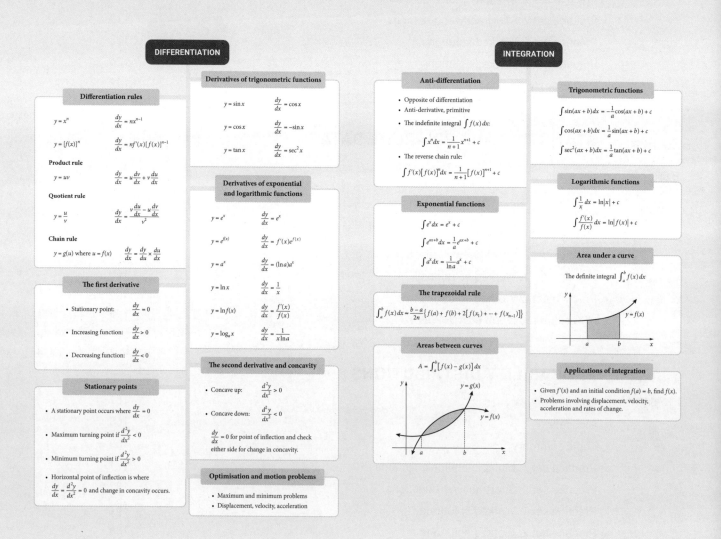

DIFFERENTIATION

Differentiation rules

$y = x^n \qquad \dfrac{dy}{dx} = nx^{n-1}$

$y = [f(x)]^n \qquad \dfrac{dy}{dx} = nf'(x)[f(x)]^{n-1}$

Product rule

$y = uv \qquad \dfrac{dy}{dx} = u\dfrac{dv}{dx} + v\dfrac{du}{dx}$

Quotient rule

$y = \dfrac{u}{v} \qquad \dfrac{dy}{dx} = \dfrac{v\dfrac{du}{dx} - u\dfrac{dv}{dx}}{v^2}$

Chain rule

$y = g(u)$ where $u = f(x) \qquad \dfrac{dy}{dx} = \dfrac{dy}{du} \times \dfrac{du}{dx}$

The first derivative
- Stationary point: $\dfrac{dy}{dx} = 0$
- Increasing function: $\dfrac{dy}{dx} > 0$
- Decreasing function: $\dfrac{dy}{dx} < 0$

Stationary points
- A stationary point occurs where $\dfrac{dy}{dx} = 0$
- Maximum turning point if $\dfrac{d^2y}{dx^2} < 0$
- Minimum turning point if $\dfrac{d^2y}{dx^2} > 0$
- Horizontal point of inflection is where $\dfrac{dy}{dx} = \dfrac{d^2y}{dx^2} = 0$ and change in concavity occurs.

Derivatives of trigonometric functions

$y = \sin x \qquad \dfrac{dy}{dx} = \cos x$

$y = \cos x \qquad \dfrac{dy}{dx} = -\sin x$

$y = \tan x \qquad \dfrac{dy}{dx} = \sec^2 x$

Derivatives of exponential and logarithmic functions

$y = e^x \qquad \dfrac{dy}{dx} = e^x$

$y = e^{f(x)} \qquad \dfrac{dy}{dx} = f'(x)e^{f(x)}$

$y = a^x \qquad \dfrac{dy}{dx} = (\ln a)a^x$

$y = \ln x \qquad \dfrac{dy}{dx} = \dfrac{1}{x}$

$y = \ln f(x) \qquad \dfrac{dy}{dx} = \dfrac{f'(x)}{f(x)}$

$y = \log_a x \qquad \dfrac{dy}{dx} = \dfrac{1}{x \ln a}$

The second derivative and concavity
- Concave up: $\dfrac{d^2y}{dx^2} > 0$
- Concave down: $\dfrac{d^2y}{dx^2} < 0$

$\dfrac{dy}{dx} = 0$ for point of inflection and check either side for change in concavity.

Optimisation and motion problems
- Maximum and minimum problems
- Displacement, velocity, acceleration

INTEGRATION

Anti-differentiation
- Opposite of differentiation
- Anti-derivative, primitive
- The indefinite integral $\int f(x)\,dx$:

$$\int x^n\,dx = \dfrac{1}{n+1}x^{n+1} + c$$

- The reverse chain rule:

$$\int f'(x)[f(x)]^n\,dx = \dfrac{1}{n+1}[f(x)]^{n+1} + c$$

Exponential functions

$\int e^x\,dx = e^x + c$

$\int e^{ax+b}\,dx = \dfrac{1}{a}e^{ax+b} + c$

$\int a^x\,dx = \dfrac{1}{\ln a}a^x + c$

The trapezoidal rule

$$\int_a^b f(x)\,dx \approx \dfrac{b-a}{2n}\left\{f(a) + f(b) + 2[f(x_1) + \cdots + f(x_{n-1})]\right\}$$

Areas between curves

$$A = \int_a^b [f(x) - g(x)]\,dx$$

Trigonometric functions

$\int \sin(ax + b)\,dx = -\dfrac{1}{a}\cos(ax + b) + c$

$\int \cos(ax + b)\,dx = \dfrac{1}{a}\sin(ax + b) + c$

$\int \sec^2(ax + b)\,dx = \dfrac{1}{a}\tan(ax + b) + c$

Logarithmic functions

$\int \dfrac{1}{x}\,dx = \ln|x| + c$

$\int \dfrac{f'(x)}{f(x)}\,dx = \ln|f(x)| + c$

Area under a curve

The definite integral $\int_a^b f(x)\,dx$

Applications of integration
- Given $f'(x)$ and an initial condition $f(a) = b$, find $f(x)$.
- Problems involving displacement, velocity, acceleration and rates of change.

SERIES, INVESTMENTS, LOANS AND ANNUITIES

📎 Common content with the Mathematics Standard 2 course

Arithmetic sequences and series

$$T_n = a + (n-1)d$$
$$T_n = T_{n-1} + d$$
$$S_n = \frac{n}{2}[2a + (n-1)d]$$
$$S_n = \frac{n}{2}(a + l)$$

Investments

📎 • Compound interest
• Effective interest rate

Annuities

📎 • Future value
📎 • Present value
📎 • By tables and recurrence relations
• By geometric series

Geometric sequences and series

$$T_n = ar^{n-1}$$
$$T_n = rT_{n-1}$$
$$S_n = \frac{a(r^n - 1)}{r - 1}$$
or $$S_n = \frac{a(1 - r^n)}{1 - r}$$
$$S_\infty = \frac{a}{1 - r}, \quad |r| < 1$$

Reducing balance loans

📎 • By tables and recurrence relations
• By geometric series

STATISTICS AND BIVARIATE DATA

📎 Other than variance, this entire topic is common content with the Mathematics Standard 2 course.

Types of data

• Categorical: nominal/ordinal
• Numerical: discrete/continuous

Measures of spread

• Range
• Quartiles, deciles and percentiles
• Interquartile range (IQR) $= Q_3 - Q_1$
• Variance, σ^2
• Standard deviation, σ
• Sample standard deviation, s
• An outlier is below $Q_1 - 1.5 \times$ IQR or above $Q_3 + 1.5 \times$ IQR

Scatterplots

Used to graph bivariate data (2 variables)

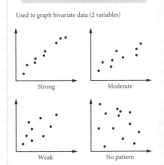

Strong   Moderate

Weak   No pattern

Line of best fit

• Dependent and independent variables
• Drawing by eye
• Using technology: least-squares regression line
• Interpolation and extrapolation

Measures of central tendency

• Mean, $\bar{x} = \dfrac{\text{sum of values}}{\text{number of values}}$
  $\bar{x} = \dfrac{\Sigma fx}{\Sigma f}$ for fx table
• Median: middle value
• Mode: most common value

The shape of a statistical distribution.

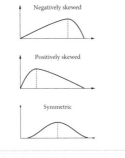

Negatively skewed

Positively skewed

Symmetric

Correlation

• Pearson's correlation coefficient, r, where $-1 \le r \le 1$ for linear relationships:
  $r = -1$:   strong, negative
  $r = -0.5$:   moderate, negative
  $r = 0$:   no correlation
  $r = 0.5$:   moderate, positive
  $r = 1$:   strong, positive

PROBABILITY DISTRIBUTIONS

📎 Common content with the Mathematics Standard 2 course

Continuous probability distributions

Probability density function (PDF)

$$\int_{-\infty}^{\infty} f(x)\,dx = 1$$

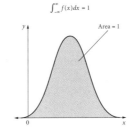

Area = 1

$$P(a \le X \le b) = \int_a^b f(x)\,dx$$

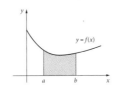

$y = f(x)$

Cumulative distribution function (CDF)

$$F(x) = \int_a^x f(x)\,dx$$

Quantiles

• Median: $\int_a^x f(x)\,dx = \frac{1}{2}$
• Lower quartile: $\int_a^x f(x)\,dx = \frac{1}{4}$
• Upper quartile: $\int_a^x f(x)\,dx = \frac{3}{4}$
• Deciles: 3rd decile > bottom 30% of values
  $\rightarrow \int_a^x f(x)\,dx = 0.3$
• Percentiles: 64th percentile > bottom 64% of values
  $\rightarrow \int_a^x f(x)\,dx = 0.64$

Normal distribution 📎

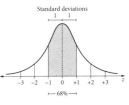

Mean
Mode
Median

z-scores 📎

• Measures number of standard deviations from the mean: $z = -1.6$ means 1.6 standard deviations below the mean.
$$z = \frac{x - \mu}{\sigma}$$

• **Empirical rule**

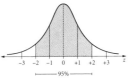

Standard deviations

68%

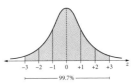

95%

99.7%

• Probability tables for z-scores
• Comparing z-scores

SYLLABUS REFERENCE GRID

Topic and subtopics	A+ HSC Year 12 Mathematics Advanced Study Notes chapter
FUNCTIONS	
MA-F2 Graphing techniques	1 Graphing functions
TRIGONOMETRIC FUNCTIONS	
MA-T3 Trigonometric functions and graphs	2 Trigonometric functions
CALCULUS	
MA-C2 Differential calculus C2.1 Differentiation of trigonometric, exponential and logarithmic functions C2.2 Rules of differentiation	3 Differentiation
MA-C3 Applications of differentiation C3.1 The first and second derivatives C3.2 Applications of the derivative	3 Differentiation
MA-C4 Integral calculus C4.1 The anti-derivative C4.2 Areas and the definite integral	4 Integration
FINANCIAL MATHEMATICS	
MA-M1 Modelling financial situations M1.1 Modelling investments and loans M1.2 Arithmetic sequences and series M1.3 Geometric sequences and series M1.4 Financial applications of sequences and series	5 Series, investments, loans and annuities
STATISTICAL ANALYSIS	
MA-S2 Descriptive statistics and bivariate data analysis S2.1 Data (grouped and ungrouped) and summary statistics S2.2 Bivariate data analysis	6 Statistics and bivariate data
MA-S3 Random variables S3.1 Continuous random variables S3.2 The normal distribution	7 Probability distributions

ABOUT THE AUTHORS

Sarah Hamper is the Head Teacher of Mathematics at Cheltenham Girls' High School and has taught for over 20 years, mostly at Abbotsleigh School, Wahroonga. Sarah co-wrote *New Century Maths 11–12 Mathematics Standard 2* and *New Century Maths 9–10* and *9–10 Advanced*.

A+ DIGITAL FLASHCARDS

Revise key terms and concepts online with the A+ Flashcards. Each topic glossary in this book has a corresponding deck of digital flashcards you can use to test your understanding and recall. Just scan the QR code or type the URL into your browser to access them.

Note: You will need to create a free NelsonNet account.

https://get.ga/a-hsc-maths-advanced

HSC EXAM FORMAT

Mathematics Advanced HSC exam

The information below about the Mathematics Advanced HSC exam was correct at the time of printing in 2021. Please check the NESA website in case it has changed.

Visit www.educationstandards.nsw.edu.au, select 'Year 11 – Year 12', 'Syllabuses A–Z', 'Mathematics Advanced', then 'Assessment and reporting in Mathematics Advanced Stage 6'. Scroll down to the heading 'HSC examination specifications'.

	Questions	Marks	Recommended time
Section I	10 multiple-choice questions Mark answers on the multiple-choice answer sheet.	10	15 min
Section II	Approx. 21 short-answer questions, including 2 or more questions worth 4 or 5 marks. Write answers on the lines provided on the paper.	90	2 h 45 min
Total		100	3 h

Exam information and tips

- Reading time: 10 minutes; use this time to preview the whole exam.

- Working time: 3 hours

- Questions focus on Year 12 outcomes but Year 11 knowledge may be examined.

- Answers are to be written on the question paper.

- A reference sheet is provided at the back of the exam paper, containing common formulas.

- Common questions with the Mathematics Standard 2 HSC exam: 20–25 marks

- The 4- or 5-mark questions are usually complex problems that require many steps of working and careful planning.

- To help you plan your time, the halfway point of Section II is marked by a notice at the bottom of the relevant page; for example, 'Questions 21–31 are worth 58 marks in total'.

- Having 3 hours for a total of 100 marks means that you have an average of 1.8 minutes per mark (or 5 minutes for 3 marks).

- If you budget 15 minutes for Section I and 1 hour 15 minutes for each half of Section II, then you will have 15 minutes at the end of the exam to check over your work and complete questions you missed.

STUDY AND EXAM ADVICE

A journey of a thousand miles begins with a single step.

Lao Tzu (c. 570–490 BCE), Chinese philosopher

I've always believed that if you put in the work, the results will come.

Michael Jordan (1963–), American basketball player

Four PRACtical steps for maths study

1. **P**ractise your maths

- Do your homework.
- Learning maths is about mastering a collection of skills.
- You become successful at maths by doing it more, through regular practice and learning.
- Aim to achieve a high level of understanding.

2. **R**ewrite your maths

- Homework and study are not the same thing. Study is your private 'revision' work for strengthening your understanding of a subject.
- Before you begin any questions, make sure you have a thorough understanding of the topic.
- Take ownership of your maths. Rewrite the theory and examples in your own words.
- Summarise each topic to see the 'whole picture' and know it all.

3. **A**ttack your maths

- All maths knowledge is interconnected. If you don't understand one topic fully, then you may have trouble learning another topic.
- Mathematics is not an HSC course you can learn 'by halves' – you have to know it all!
- Fill in any gaps in your mathematical knowledge to see the 'whole picture'.
- Identify your areas of weakness and work on them.
- Spend most of your study time on the topics you find difficult.

4. **C**heck your maths

- After you have mastered a maths skill, such as graphing a quadratic equation, no further learning or reading is needed, just more practice.
- Compared to other subjects, the types of questions asked in maths exams are conventional and predictable.
- Test your understanding with revision exercises, practice papers and past exam papers.
- Develop your exam technique and problem-solving skills.
- Go back to steps 1–3 to improve your study habits.

9780170459228

Topic summaries and concept maps

Summarise each topic when you have completed it, to create useful study notes for revising the course, especially before exams. Use a notebook or folder to list the important ideas, formulas, terminology and skills for each topic. This book is a good study guide, but educational research shows that effective learning takes place when you rewrite learned knowledge in your own words.

A good topic summary runs for 2 to 4 pages. It is a condensed, personalised version of your course notes. This is your interpretation of a topic, so include your own comments, symbols, diagrams, observations and reminders. Highlight important facts using boxes and include a glossary of key words and phrases.

A concept map or mind map is a topic summary in graphic form, with boxes, branches and arrows showing the connections between the main ideas of the topic. This book contains examples of concept maps. The topic name is central to the map, with key concepts or subheadings listing important details and formulas. Concept maps are powerful because they present an overview of a topic on one large sheet of paper. Visual learners absorb and recall information better using concept maps.

When compiling a topic summary, use your class notes, your textbook and this study guide. Ask your teacher for a copy of the course syllabus or the school's teaching program, which includes the objectives and outcomes of every topic in dot point form.

Attacking your weak areas

Most of your study time should be spent on attacking your weak areas to fill in any gaps in your maths knowledge. Don't spend too much time on work you already know well, unless you need a confidence boost! Ask your teacher, use this book or your textbook to improve the understanding of your weak areas and to practise maths skills. Use your topic summaries for general revision, but spend longer study periods on overcoming any difficulties in your mastery of the course.

Practising with past exam papers

Why is practising with past exam papers such an effective study technique? It allows you to become familiar with the format, style and level of difficulty expected in an HSC exam, as well as the common topic areas tested. Knowing what to expect helps alleviate exam anxiety. Remember, mathematics is a subject in which the exam questions are fairly predictable. The exam writers are not going to ask too many unusual questions. By the time you have worked through many past exam papers, this year's HSC paper won't seem that much different.

Don't throw your old exam papers away. Use them to identify your mistakes and weak areas for further study. Revising topics and then working on mixed questions is a great way to study maths. You might like to complete a past HSC exam paper under timed conditions to improve your exam technique.

Past HSC exam papers are available at the NESA website: visit www.educationstandards.nsw.edu.au and select 'Year 11 – Year 12', 'HSC exam papers'. NESA marking feedback and guidelines can also be viewed there. Cengage has also published *A+ HSC Year 12 Mathematics Advanced Practice Exams*, containing topic exams and practice HSC exam papers. You can find past HSC exam papers with solutions online, in bookstores, at the Mathematical Association of NSW (www.mansw.nsw.edu.au) and at your school (ask your teacher) or library.

Preparing for an exam

- Make a study plan early; don't leave it until the last minute.

- Read and revise your topic summaries.

- Work on your weak areas and learn from your mistakes.

- Don't spend too much time studying work you know already.

- Revise by completing revision exercises and past exam papers or assignments.

- Vary the way you study so that you don't become bored: ask someone to quiz you, voice-record your summary, design a poster or concept map, or explain the work to someone.

- Anticipate the exam:
  - How many questions will there be?
  - What are the types of questions: multiple-choice, short-answer, long-answer, problem-solving?
  - Which topics will be tested?
  - How many marks are there in each section?
  - How long is the exam?
  - How much time should I spend on each question/section?
  - Which formulas are on the reference sheet and how do I use them in the exam?

During an exam

1. Bring all of your equipment, including a ruler and calculator (check that your calculator works and is in RADIANS mode for trigonometric functions, but use DEGREES for trigonometric measurements). A highlighter pen may help for tables, graphs and diagrams.

2. Don't worry if you feel nervous before an exam – this is normal and helps you perform better; however, being too casual or too anxious can harm your performance. Just before the exam begins, take deep, slow breaths to reduce any stress.

3. Write clearly and neatly in black or blue pen, not red. Use a pencil only for diagrams and constructions.

4. Use the **reading time** to browse through the exam to see the work that is ahead of you and the marks allocated to each question. Doing this will ensure you won't miss any questions or pages. Note the harder questions and allow more time for working on them. Leave them if you get stuck, and come back to them later.

5. Attempt every question. It is better to do most of every question and score some marks, rather than ignore questions completely and score 0 for them. Don't leave multiple-choice questions unanswered! Even if you guess, you have a chance of being correct.

6. Easier questions are usually at the beginning, with harder ones at the end. Do an easy question first to boost your confidence. Some students like to leave multiple-choice questions until last so that, if they run out of time, they can make quick guesses. However, some multiple-choice questions can be quite difficult.

7. Read each question and identify what needs to be found and what topic/skill it is testing. The number of marks indicates how much time and working out is required. Highlight any important keywords or clues. Do you need to use the answer to the previous part of the question?

8. After reading each question, and before you start writing, spend a few moments planning and thinking.

9. You don't need to be writing all of the time. What you are writing may be wrong and a waste of time. Spend some time considering the best approach.

10. Make sure each answer seems reasonable and realistic, especially if it involves money or measurement.

11. Show all necessary working, write clearly, draw big diagrams, and set your working out neatly. Write solutions to each part underneath the previous step so that your working out goes down the page, not across.

12. Use a ruler to draw (or read) half-page graphs with labels and axes marked, or to measure scale diagrams.

13. Don't spend too much time on one question. Keep an eye on the time.

14. Make sure you have answered the question. Did you remember to round the answer and/or include units? Did you use all of the relevant information given?

15. If a hard question is taking too long, don't get bogged down. If you're getting nowhere, retrace your steps, start again, or skip the question (circle it) and return to it later with a clearer mind.

16. If you make a mistake, cross it out with a neat line. Don't scribble over it completely or use correction fluid or tape (which is time-consuming and messy). You may still score marks for crossed-out work if it is correct, but don't leave multiple answers! Keep track of your answer booklets and ask for more writing paper if needed.

17. Don't cross out or change an answer too quickly. Research shows that often your first answer is the correct one.

18. Don't round your answer in the middle of a calculation. Round at the end only.

19. Be prepared to write words and sentences in your answers, but don't use abbreviations that you've just made up. Use correct terminology and write 1 or 2 sentences for 2 or 3 marks, not mini-essays.

20. If you have time at the end of the exam, double-check your answers, especially for the more difficult or uncertain questions.

Ten exam habits of the best HSC students

1. Has clear and careful working and checks their answers

2. Has a strong understanding of basic algebra and calculation

3. Reads (and answers) the whole question

4. Chooses the simplest and quickest method

5. Checks that their answer makes sense or sounds reasonable

6. Draws big, clear diagrams with details and labels

7. Uses a ruler for drawing, measuring and reading graphs

8. Can explain answers in words when needed, in 1−2 clear sentences

9. Uses the previous parts of a question to solve the next part of the question

10. Rounds answers at the end, not before

Further resources

Visit the NESA website, www.educationstandards.nsw.edu.au, for the following resources.
Select 'Year 11 − Year 12' and then 'Syllabuses A−Z' or 'HSC exam papers'.

- Mathematics Advanced Syllabus

- Past HSC exam papers, including marking feedback and guidelines

- Sample HSC questions/exam papers and marking guidelines

Before 2020, 'Mathematics Advanced' was called simply 'Mathematics'. For these exam papers, select 'Year 11 − Year 12', 'Resources archive', 'HSC exam papers archive'.

MATHEMATICAL VERBS

A glossary of 'doing words' common in maths problems and HSC exams

analyse
study in detail the parts of a situation

apply
use knowledge or a procedure in a given situation

calculate
See **evaluate**

classify/identify
state the type, name or feature of an item or situation

comment
express an observation or opinion about a result

compare
show how two or more things are similar or different

complete
fill in detail to make a statement, diagram or table correct or finished

construct
draw an accurate diagram

convert
change from one form to another, for example, from a fraction to a decimal, or from kilograms to grams

decrease
make smaller

describe
state the features of a situation

estimate
make an educated guess for a number, measurement or solution, to find roughly or approximately

evaluate/calculate
find the value of a numerical expression, for example, 3×8^2 or $4x + 1$ when $x = 5$

expand
remove brackets in an algebraic expression, for example, expanding $3(2y + 1)$ gives $6y + 3$

explain
describe why or how

give reasons
show the rules or thinking used when solving a problem. *See also* **justify**

graph
display on a number line, number plane or statistical graph

hence find/prove
calculate an answer or prove a result using previous answers or information supplied

identify
See **classify**

increase
make larger

interpret
find meaning in a mathematical result

justify
give reasons or evidence to support your argument or conclusion. *See also* **give reasons**

measure
determine the size of something, for example, using a ruler to determine the length of a pen

prove
See **show/prove that**

recall
remember and state

show/prove that
(in questions where the answer is given) use calculation, procedure or reasoning to prove that an answer or result is true

simplify
express a result such as a ratio or algebraic expression in its most basic, shortest, neatest form

sketch
draw a rough diagram that shows the general shape or ideas (less accurate than **construct**)

solve
calculate the value(s) of an unknown pronumeral in an equation or inequality

state
See **write**

substitute
replace part of an expression with another, equivalent expression.

verify
check that a solution or result is correct, usually by substituting back into an equation or referring back to the problem

write/state
give an answer, formula or result without showing any working or explanation (This usually means that the answer can be found mentally, or in one step)

9780170459228

SYMBOLS AND ABBREVIATIONS

$=$	is equal to	$[a, b], a \le x \le b$	the interval of x-values from a to b (including a and b)
$\ne$	is not equal to	$(a, b), a < x < b$	the interval of x-values between a and b (excluding a and b)
$\approx$	is approximately equal to		
$<$	is less than	S 37° W	a compass bearing
$>$	is greater than	217°	a true bearing
$\le$	is less than or equal to	$P(E)$	the probability of event E occurring
$\ge$	is greater than or equal to	$P(\bar{E})$	the probability of event E not occurring
( )	parentheses, round brackets		
[ ]	(square) brackets	$P(A \mid B)$	the probability of A given B
{ }	braces	$A \cup B$	A union B, A or B
$\pm$	plus or minus	$A \cap B$	A intersection B, A and B
π	pi = 3.141 59...	PDF	probability density function
$\equiv$	is congruent/identical to	CDF	cumulative distribution function
$0.1\dot{5}\dot{2}$	the recurring decimal 0.152 152...	LHS	left-hand side
°	degree	RHS	right-hand side
$\angle$	angle	p.a.	per annum (per year)
Δ	triangle	$\bar{x}$	the mean
$\parallel$	is parallel to	$\mu = E(X)$	the population mean, expected value
$\perp$	is perpendicular to		
$\therefore$	therefore	σ_n	the standard deviation
$\sqrt{}$	square root	$\text{Var}(X) = \sigma^2$	the variance
$\sqrt[3]{}$	cube root	Σ	the sum of, sigma
∞	infinity	Q_1	first quartile or lower quartile
$\lvert x \rvert$	absolute value of x	Q_2	median (second quartile)
$\lim\limits_{h \to 0}$	the limit as $h \to 0$	Q_3	third quartile or upper quartile
		IQR	interquartile range
$\dfrac{dy}{dx}, y', f'(x)$	the 1st derivative of $y, f(x)$	α	alpha
		θ	theta
$\dfrac{d^2y}{dx^2}, y'', f''(x)$	the 2nd derivative of $y, f(x)$	m	gradient
$\int f(x)\,dx$	the integral of $f(x)$		

A+ HSC YEAR 12 MATHEMATICS

STUDY NOTES

 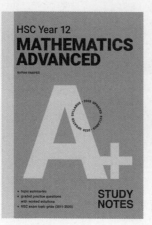  

Authors:

Tania Eastcott   Sarah Hamper   Karen Man   Jim Green
Rachel Eastcott   Ashleigh Della Marta   Janet Hunter

PRACTICE EXAMS

Authors:

Adrian Kruse   Simon Meli   John Drake   Jim Green
                                                                     Janet Hunter

9780170459228

CHAPTER 1
GRAPHING FUNCTIONS

9780170459228

GRAPHING FUNCTIONS

Functions

- Polynomial functions
- Reciprocal functions
- Absolute value functions
- Exponential functions
- Logarithmic functions
- Domain and range
- Horizontal and vertical asymptotes
- x- and y-intercepts
- Applications of functions

Transformation of functions

- Translations
- Dilations
- Reflections in the x- and y-axes
- $y = kf(a(x + b)) + c$
- Combined transformations
- Domain and range
- Asymptotes and intercepts
- Symmetry and discontinuities

Solving equations and inequalities graphically

- Solving equations graphically or counting the number of solutions
- Solving linear and quadratic inequalities graphically

Glossary

asymptote

A line in which a curve approaches but never touches. For this exponential curve, the x-axis is an asymptote.

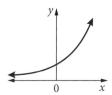

A **horizontal asymptote** exists when, for very large values of x (in either the positive or negative direction), the values of y approaches a constant (b in the following diagram).

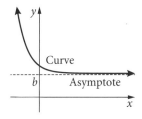

A **vertical asymptote** exists where a function is not defined for a specific value of x (c in the diagram) but the values of y become very large in size (positive or negative) as x approaches it.

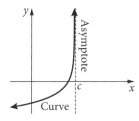

dilation

A transformation of a function that results in its graph being stretched or compressed horizontally or vertically.

domain

The set of x-values for which the function $y = f(x)$ is defined; the 'input' values of a function, the values of the independent variable.

function

A relation where each input is mapped to a single output. For the function $y = f(x)$, each value of x is mapped to exactly one (unique) value of y, where x is the independent variable and y is the dependent variable.

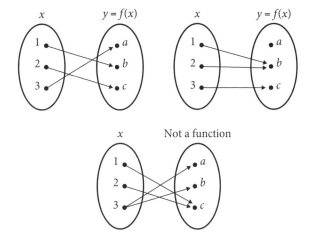

range

The set of y-values for which the function $y = f(x)$ is defined; the 'output' values of a function, the values of the dependent variable.

reflection

A transformation of a function that results in its graph being flipped horizontally ('back-to-front') or vertically ('upside-down').

scale factor

The value (k) by which the graph of a function is dilated.

sketch

To draw a function that shows important features, such as intercepts and asymptotes, but the graph is not drawn precisely to scale.

transformation

The modification of a function and its graph by translating, reflecting, stretching or compressing it.

translation

A transformation of a function that results in its graph being horizontally and/or vertically shifted. Its size and shape remain the same.

Topic summary

Graphing techniques (MA-F2)

Vertical translations of functions

For the function $y = f(x)$, $y = f(x) + c$ is a vertical translation.

If $c > 0$, the graph is translated upwards by c units.
If $c < 0$, the graph is translated downwards by c units.

'c' shifts the function up or down *without* changing the size or shape of its graph. It changes the y-values of the function.

For example, $y = \sin x + 2$ is the function $y = \sin x$ translated up 2 units.

Note: Transformations of trigonometric functions are covered in Chapter 2. See Example 6 on page 32 for the graph of $y = \sin x + 2$.

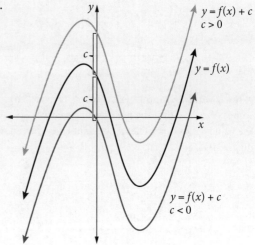

Horizontal translations of functions

For the function $y = f(x)$, $y = f(x + b)$ is a horizontal translation.

If $b > 0$, the graph is translated left by b units.
If $b < 0$, the graph is translated right by b units.

'b' shifts the function left or right without changing the size or shape of its graph. It shifts the x-values of the function.

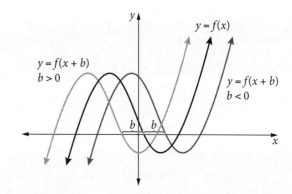

For example, $y = (x + 4)^2$ is the function $y = x^2$ translated left 4 units.

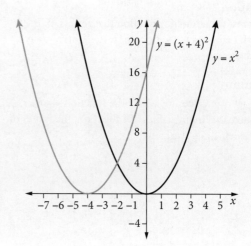

Vertical dilations of functions

For the function $y = f(x)$, $y = kf(x)$ is a vertical dilation (in the y-direction) by a scale factor of k.

If $k > 1$, the graph is stretched or expanded.
If $0 < k < 1$, the graph is compressed.

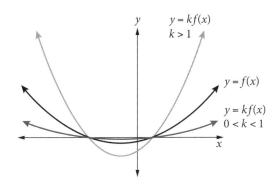

'k' dilates the function vertically and changes the size and shape of its graph. It changes the y-values of the function.

For example, $y = 3\sin x$ is the function $y = \sin x$ dilated vertically by a scale factor of 3.

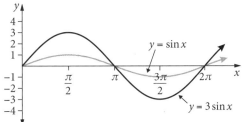

Reflections in the x-axis

$y = -f(x)$ is a reflection of $y = f(x)$ in the x-axis.

It is also a vertical dilation with scale factor $k = -1$.

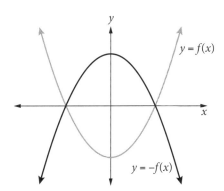

The '$-$' flips the function across the x-axis. It changes the sign of the y-values of the function.

For example, $y = -x^2$ is the function $y = x^2$ reflected in the x-axis.

Horizontal dilations of functions

For the function $y = f(x)$, $y = f(ax)$ is a horizontal dilation (in the x-direction) by a scale factor of $\dfrac{1}{a}$.

If $a > 1$, the graph is compressed.
If $0 < a < 1$, the graph is stretched.

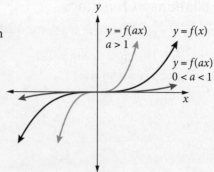

'a' dilates the function horizontally and changes the size and shape of its graph. It changes the x-values of the function. The higher the value of a, the more the graph is compressed in the x-direction from left and right.

For example, $y = \cos 2x$ is the function $y = \cos x$ dilated horizontally by a scale factor of $\dfrac{1}{2}$.

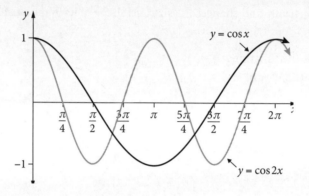

Reflections in the y-axis

$y = f(-x)$ is a reflection of $y = f(x)$ in the y-axis.

It is also a horizontal dilation by the scale factor $a = -1$.

The '$-$' flips the function across the y-axis. It changes the sign of the x-values of the function.

For example, $y = 2^{-x}$ is the function $y = 2^{x}$ reflected in the y-axis.

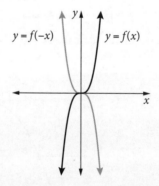

Summary of transformations

Function	Transformation to $f(x)$	Change of coordinates
$f(x) + c$	Vertical translation (up/down) of c units	$(x, y) \rightarrow (x, y + c)$
$f(x + b)$	Horizontal translation (left/right) of b units	$(x, y) \rightarrow (x + b, y)$
$-f(x)$	Reflection in the x-axis	$(x, y) \rightarrow (x, -y)$
$f(-x)$	Reflection in the y-axis	$(x, y) \rightarrow (-x, y)$
$kf(x)$	Vertical dilation by scale factor k	$(x, y) \rightarrow (x, ky)$
$f(ax)$	Horizontal dilation by scale factor $\dfrac{1}{a}$	$(x, y) \rightarrow (ax, y)$

9780170459228

Combining these, we have the general equation of a transformed function:

$$y = kf(a(x + b)) + c,$$

where $f(x)$ is a polynomial, reciprocal, absolute value, exponential or logarithmic function and a, b, c and k are constants, representing:

- a horizontal dilation of scale factor $\dfrac{1}{a}$

- a horizontal translation of b

- a vertical dilation of k

- a vertical translation of c.

For combined transformations, the above formula applies only if a dilation in one direction occurs *before* the translation in the same direction; for example,

- horizontal dilation (a), then horizontal translation (b)

- vertical dilation (k), then vertical translation (c).

Solving equations and inequalities graphically

Example 1

Solve $x^2 - 2x - 3 > 0$.

Solution

Graph the parabola $y = x^2 - 2x - 3$.

First, find x-intercepts ($y = 0$):

$$x^2 - 2x - 3 = 0$$
$$(x - 3)(x + 1) = 0$$

$$x = 3, -1$$

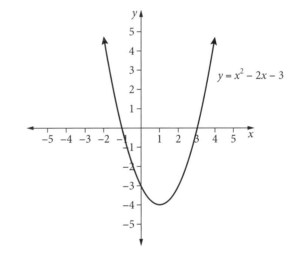

Use the above to find where $x^2 - 2x - 3 > 0$.

From the graph, the solution to $x^2 - 2x - 3 > 0$ is $x < -1$ and $x > 3$.

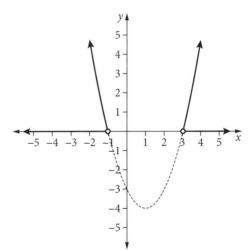

Example 2

Solve $6x + 16 - x^2 \geq 0$.

Solution

Graph $y = -x^2 + 6x + 16$.

Factorise to find the x-intercepts:

$-x^2 + 6x + 16 = 0$
$-(x^2 - 6x - 16) = 0$
$-(x - 8)(x + 2) = 0$

$x = 8, -2$

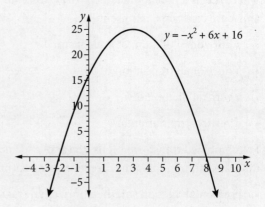

From the graph above, find where $-x^2 + 6x + 16 \geq 0$.

The solution to $-x^2 + 6x + 16 \geq 0$ is $-2 \leq x \leq 8$.

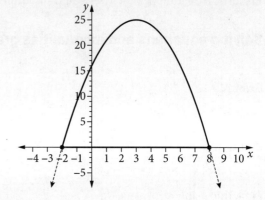

Practice set 1

Multiple-choice questions

Solutions start on page 20.

Question 1

The graph of $y = f(x)$ is shown.

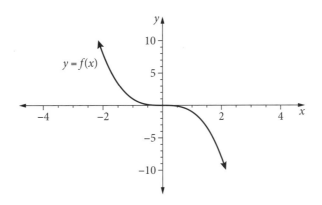

Which of the following is the graph of $y = f(x) - 2$?

A

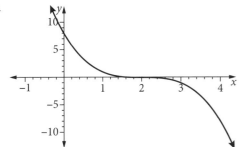

B

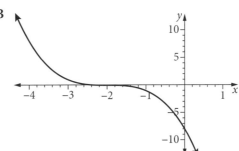

C

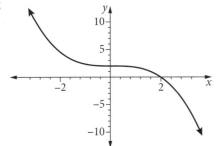

D

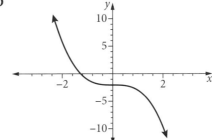

Question 2

If the point $(6, 2)$ is translated 4 units left and then reflected in the x-axis, what are its new coordinates?

A $(2, -2)$

B $(0, 4)$

C $(-10, 2)$

D $(-6, -2)$

Question 3 ⬤◯◯

The graph of $y = \sqrt{x}$ is shown below.

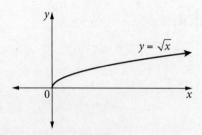

Which of the following is the graph of $y = \sqrt{x + 2}$?

A

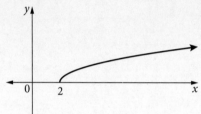

B

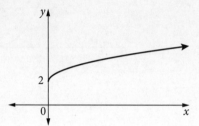

C

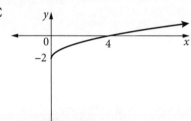

D

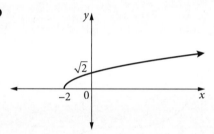

Question 4 ⬤◯◯

Which sequence of transformations changes the graph from $y = x^3$ to $y = (x - 5)^3 + 2$?

A Horizontal translation 5 units to the right and a vertical translation 2 units up

B Horizontal translation 5 units to the left and a vertical translation 2 units up

C Horizontal translation 2 units to the left and a vertical translation 5 units down

D Horizontal translation 2 units to the right and a vertical translation 5 units down

Question 5 ©NESA 2020 HSC EXAM, QUESTION 1 ⬤⬤◯

Which inequality gives the domain of $y = \sqrt{2x - 3}$?

A $x < \dfrac{3}{2}$

B $x > \dfrac{3}{2}$

C $x \le \dfrac{3}{2}$

D $x \ge \dfrac{3}{2}$

Question 6 ⬤⬤◯

The graph of the function $y = f(x)$ passes through the point $(-3, 8)$. If $g(x) = f\left(\dfrac{x}{3}\right) + 4$, then through which of the following points does the graph of $y = g(x)$ pass?

A $(-1, 12)$

B $(-1, 28)$

C $(-9, 12)$

D $\left(1, \dfrac{8}{3}\right)$

Question 7 ⬤⬤○

The function $f(x) = |x|$ is transformed and the graph of the new function is shown below.

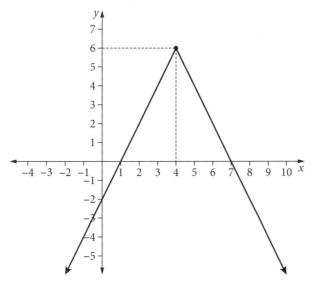

The equation of the new function is $y = k|x + b| + c$. What are the values of k, b and c?

A $k = 2$, $b = -6$ and $c = 4$

B $k = -2$, $b = -4$ and $c = 6$

C $k = 2$, $b = 4$ and $c = -6$

D $k = -2$, $b = 6$ and $c = 4$

Question 8 ⬤⬤○

The graph of $y = \dfrac{5}{x + 3}$ is translated 4 units right and dilated vertically by a factor of $\dfrac{1}{2}$.

What is the equation of the transformed graph?

A $y = \dfrac{10}{x - 1}$

B $y = \dfrac{5}{2(x - 7)}$

C $y = \dfrac{5}{2(x - 1)}$

D $y = \dfrac{10}{x - 7}$

Question 9 ⬤⬤○

Which one of the following graphs could represent the transformation of $f(x) = \log_3 x$ to the graph of $f(x) = \log_3(x - p) + q$, where p and q are both positive?

A

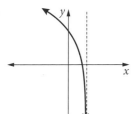

B

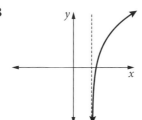

C

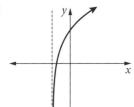

D

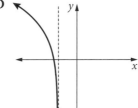

Question 10 ○●●

The graph of $y = f(x)$ is shown.

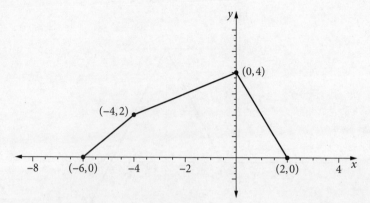

The graph of $y = f(x)$ is transformed to $y = g(x)$ as shown.

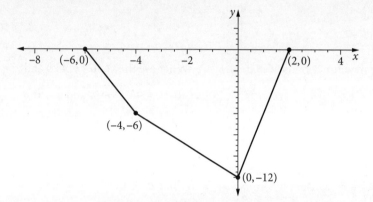

Which of the following equations describes the relationship between $g(x)$ and $f(x)$?

A $g(x) = 3f(-x)$ 　　　　**B** $g(x) = -3f(x)$ 　　　　**C** $g(x) = -f\left(\dfrac{x}{3}\right)$ 　　　　**D** $g(x) = f\left(-\dfrac{x}{3}\right)$

Question 11 ©NESA 2020 HSC EXAM, QUESTION 2 ○●●

The function $f(x) = x^3$ is transformed to $g(x) = (x - 2)^3 + 5$ by a horizontal translation of 2 units followed by a vertical translation of 5 units.

Which row of the table shows the directions of the translations?

	Horizontal translation of 2 units	Vertical translation of 5 units
A	Left	Up
B	Right	Up
C	Left	Down
D	Right	Down

Question 12 ●●●

The graph of $y = f(x)$ is shown.

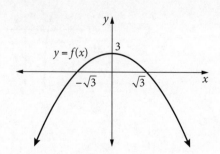

What is the graph of $y = 2f(x)$?

A

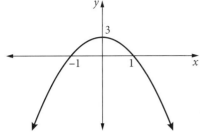

B

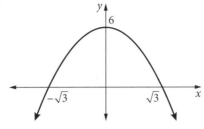

C

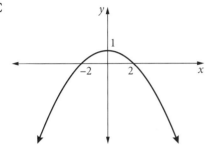

D

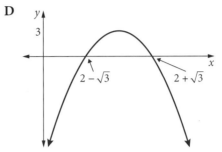

Question 13

The graph of $y = f(x)$ is shown.

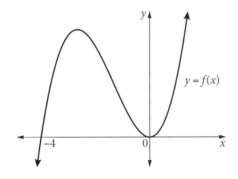

Which of the following shows the graph of $y = f(x + 1) + 1$?

A

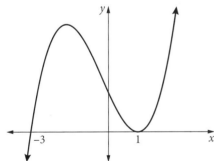

B

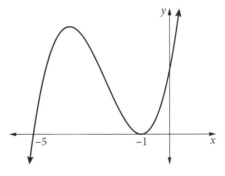

C

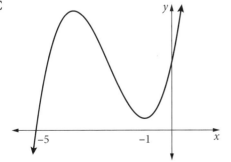

D

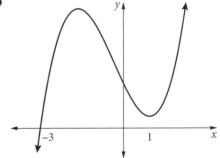

Question 14 🔘🔘⚫

The function $y = \log_e x$ is transformed by first being dilated vertically by a scale factor of 4 and then translated horizontally 3 units to the left. What is the equation of the transformed function?

A $y = 4\log_e x^3$

B $y = 3\log_e x^4$

C $y = 4\log_e(x - 3)$

D $y = 4\log_e(x + 3)$

Question 15 🔘🔘⚫

$f(x) = (x - 1)^2$ is transformed to $g(x) = 2f(x + 4) - 5$.

What is the y-intercept of the transformed function?

A -5

B -3

C 3

D 13

Question 16 🔘🔘🔘

The graph of $y = \cos x$ is shown below for the domain $\left[0, \dfrac{\pi}{2}\right]$.

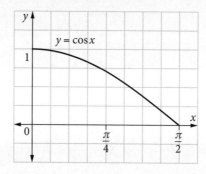

Which equation, when graphed with $y = \cos x$ for this domain, would intersect it at exactly 2 points?

A $y = |x - 2|$

B $y = |x - 1|$

C $y = 1 - |x - 1|$

D $y = 1 - |x - 2|$

Question 17 🔘🔘⚫

The graph of $y = x^3 + x$ is reflected in the y-axis, then translated 1 unit to the right.

What is the equation of the new graph?

A $y = -(x + 1)^3 - (x + 1)$

B $y = (x + 1)^3 + x + 1$

C $y = -(x - 1)^3 - (x - 1)$

D $y = (x - 1)^3 - x + 1$

Question 18 🔘🔘🔘

Which equation represents the graph of $y = 2x^3 - 11x^2 - 5x + 8$ after it has been reflected in the y-axis?

A $y = 2x^3 + 11x^2 - 5x - 8$

B $y = 2x^3 + 11x^2 + 5x - 8$

C $y = -2x^3 - 11x^2 + 5x + 8$

D $y = -2x^3 - 11x^2 - 5x + 8$

Question 19 〇〇■

The graph of $y = f(x)$ undergoes three transformations in this order:

- a vertical dilation of a units

- a horizontal translation of b units to the left

- reflection in the x-axis.

What is the equation of the transformed graph?

A $y = af(x - b)$ **B** $y = -af(x + b)$

C $y = -f(ax + b)$ **D** $y = f(ax - b)$

Question 20 〇〇〇

The diagram shows the graph of $y = e^{-x}(1 + x)$.

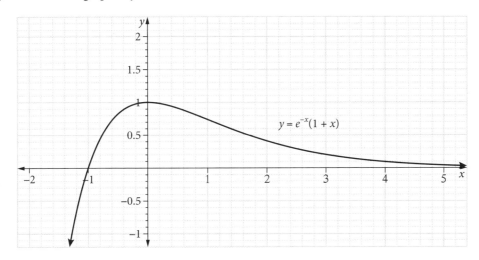

How many solutions are there to the equation $e^{-x}(1 + x) = 1 - x^2$?

A 1   **B** 2   **C** 3   **D** 4

Practice set 2

Short-answer questions

Solutions start on page 22.

Question 1 (2 marks) ⬤⬤◯

The graph of $y = f(x)$ is shown. Sketch the graph of $y = f(x) - 2$.

2 marks

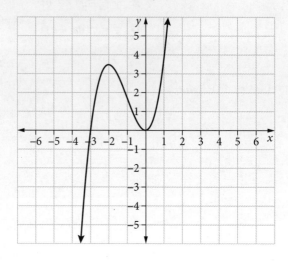

Question 2 (4 marks) ◯⬤⬤

Describe, in order, the transformations required to convert:

a $y = |x|$ to $y = 2|x - 3| + 1$

2 marks

b $y = x^3$ to $y = (x + 1)^3 - 2$

2 marks

Question 3 (3 marks) ⬤◯◯

a   Describe the transformation that would apply to $y = \sqrt{x}$ to produce $y = \sqrt{-x}$.

1 mark

b   Sketch $y = \sqrt{x}$ and $y = \sqrt{-x}$ on the same axes.

2 marks

Question 4 (2 marks) ⬤⬤◯

The function $y = 2x^2 - 4x + 3$ is dilated horizontally by a scale factor of $\frac{1}{3}$ then reflected in the y-axis. It is then translated 4 units up.

Find the equation of the transformed function.

2 marks

Question 5 (4 marks) ⬤◯◯

a   Find the y-intercept of the graph of $y = e^{-x} + 1$.

1 mark

b   Find the asymptote of the graph of $y = e^{-x} + 1$.

1 mark

c   Hence, sketch the graph of $y = e^{-x} + 1$.

2 marks

Question 6 (6 marks)

The hyperbola $y = \dfrac{1}{x+2}$ is translated horizontally 3 units to the left and then 1 unit up.

a Find the equation of the transformed graph.   2 marks

b Find all intercepts of the transformed graph.   2 marks

c Hence, sketch the transformed graph, showing all important features.   2 marks

Question 7 (3 marks)

The graph of $y = f(x)$ has a minimum point at $P(3, -4)$.

Find the coordinates of the minimum point of the curve with equation:

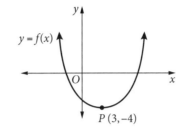

a $y = f(x-2)$ 1 mark

b $y = f(x+5) + 6$ 2 marks

Question 8 (4 marks)

The graph of $y = x^3 - 4x^2 + 1$ is shown.

a Use the graph to solve the equation $x^3 - 4x^2 + 1 = 1$.   2 marks

b Hence, state all values of k for which $x^3 - 4x^2 + 1 = k$ has two or more solutions.   2 marks

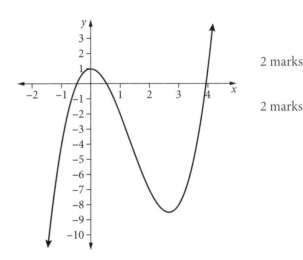

Question 9 (1 mark)

Describe the transformation that maps the graph of $y = \sqrt{8x^3 + 1}$ onto the graph of $y = \sqrt{x^3 + 1}$.   1 mark

Question 10 (3 marks)

The graph shown is of a function in the form $f(x) = k\log_e(x + a) + c$, where k, a and c are constants.

The graph has a vertical asymptote with equation $x = -1$, a y-intercept at 1 and a point $P(p, 10)$.

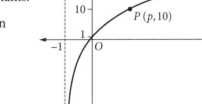

a State the value of a.   1 mark

b Determine the value of c.   1 mark

c Show that $k = \dfrac{9}{\log_e(p+1)}$.   1 mark

Question 11 (2 marks)

The function $y = f(x)$ is transformed to $y = g(x)$, where $g(x) = \dfrac{3}{2} f\left(\dfrac{1}{4}(x-3)\right) - 5$.

What are the transformed coordinates of the point $(-2, 3)$ on the graph of $y = g(x)$?   2 marks

Question 12 (5 marks)

The graph on the right shows the predicted weekly profits from the sale of face masks during a pandemic. It can be modelled by the cubic function $y = 400x^3 - 2300x^2 + 6000x + 3000$, where y is the profit in dollars and x is the number of weeks after 1 April 2020.

Due to high demand, the company was ahead in production of the face masks, and sales began one week earlier.

Projected profit from 1 April 2020

a Describe in words the transformation that can be made to the function and graph if the same profits are now being achieved one week earlier.   1 mark

b Write the equation of the transformed function. (Do not expand or simplify your equation.)   1 mark

c According to the transformed function:

   i calculate the increase in profit made between weeks 4 and 5 after 1 April 2020.   2 marks

   ii estimate the week in which a profit of $20 000 is made.   1 mark

Question 13 (6 marks)

The function $y = 2^x$ is transformed to $y = 2^{x+3} - 3$.

a Describe the transformations that have taken place.   2 marks

b Accurately sketch the graph of $y = 2^{x+3} - 3$.   2 marks

c State the domain and range of the transformed function.   2 marks

Question 14 (2 marks)

Zoe was considering the graph of the function $h(x) = -f[3(x - 6)] - 10$ and observed that the point $(4, 0)$ lies on this graph.

Where would this point be located on the original graph of $y = f(x)$?   2 marks

Question 15 (3 marks) ©NESA 2020 HSC EXAM, QUESTION 24

The circle $x^2 - 6x + y^2 + 4y - 3 = 0$ is reflected in the x-axis.

Sketch the reflected circle, showing the coordinates of the centre and the radius.   3 marks

Question 16 (2 marks) ©NESA SAMPLE 2020 HSC EXAM, QUESTION 24

The function $f(x) = |x|$ is transformed and the equation of the new function is of the form $y = kf(x + b) + c$, where k, b and c are constants.

The graph of the new function is shown.

What are the values of k, b and c?   2 marks

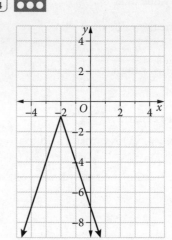

Question 17 (3 marks) ©NESA SAMPLE 2020 HSC EXAM, QUESTION 26 ●●●

By drawing graphs on the number plane, determine how many solutions there are to the   3 marks

equation $\sin x = \left| \dfrac{x}{5} \right|$ in the domain $(-\infty, \infty)$.

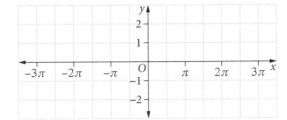

Question 18 (7 marks) ●●●

The diagram below shows the graph of a curve with equation $f(x) = \dfrac{x}{x-3}$, where $x \neq 3$.

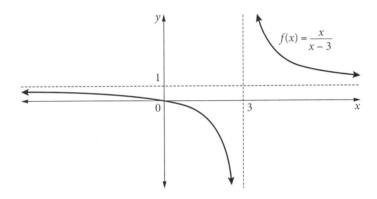

The curve passes through the origin and has 2 asymptotes, as shown in the diagram.

a State the equations of the asymptotes.   2 marks

b Sketch the graph of $y = f(x + 1)$.   3 marks

c Hence, find the coordinates of the points where the curve $y = f(x + 1)$ crosses the   2 marks
   x- and y-axes.

Question 19 (4 marks) ●●●

a Sketch the graph of $y = |2x - 5|$.   2 marks

b Using this graph, determine the value(s) of m for which the equation $|2x - 5| = mx - 1$ 2 marks
   has only 1 solution.

Question 20 (6 marks) ●●●

a On one set of axes, sketch the graphs of both $f(x) = x^2 - 4$ and $g(x) = 3x$, clearly labelling   2 marks
   the x- and y-intercepts.

b Using your graph in part **a**, find the points of intersection, where $f(x) = g(x)$.   2 marks

c Find the values of x for which $x^2 - 4 > 3x$.   2 marks

Practice set 1

Worked solutions

1 D

Vertical translation down 2 units

2 A

(6, 2)	
(6 − 4, 2)	translated left 4 units
(2, −2)	reflected in x-axis

3 D

$f(x + 2)$	translation left 2 units

$$y = \sqrt{x - (-2)}$$
$$= \sqrt{x + 2}$$

4 A

$y = x^3$

$= (x - 5)^3$	horizontal translation 5 units right
$= (x - 5)^3 + 2$	vertical translation 2 units up

5 D

$$2x - 3 \geq 0$$
$$2x \geq 3$$
$$x \geq \frac{3}{2}$$

6 C

$y = f(x)$ passes through $(-3, 8)$

$$g(x) = f\left(\frac{x}{3}\right) + 4$$

Horizontal dilation: $(-3 \times 3, 8)$
  $(-9, 8)$

Vertical translation up 4 units: $(-9, 8 + 4)$
  $(-9, 12)$

7 B

The graph of $f(x) = |x|$ has a vertex at $(0, 0)$. It has been reflected in the x-axis, so $k < 0$. The vertex has been translated 4 units right and 6 units up, so its new equation is

$f(x) = k|x - 4| + 6$.

To find k, choose a point, say $(1, 0)$:

$$f(1) = k|1 - 4| + 6 = 0$$
$$k \times 3 = -6$$
$$k = -2$$

So $k = -2$, $b = -4$, $c = 6$.

8 C

$$y = \frac{5}{x + 3}$$

Translate 4 units right:
$x + 3 = x - (-3 + 4) = x - 1$

Vertical dilation factor $\frac{1}{2}$

$$y = \frac{1}{2} \times \frac{5}{x - 1}$$
$$= \frac{5}{2(x - 1)}$$

9 B

Horizontal translation right p units
Vertical translation q units up

10 B

Consider a point:
$(0, 4)$ becomes $(0, 4 \times 3) = (0, 12)$
Reflects in x-axis $(0, -12)$

11 B

Horizontal translation of 2 units right and vertical translation of 5 units up

12 B

Vertical dilation factor of 2
For y-intercept of 3, $y = 2 \times 3 = 6$.
The x-intercepts do not change.

13 C

Horizontal translation of 1 unit left and vertical shift of 1 unit up

$(0, 0)$ becomes $(0 - 1, 0 + 1) = (-1, 1)$

$(-4, 0)$ becomes $(-4 - 1, 0 + 1) = (-5, 1)$

14 D

$y = \log_e x$
$= 4 \log_e x$
$= 4 \log_e (x - (-3))$
$= 4 \log_e (x + 3)$

15 D

$g(x) = 2(x + 3)^2 - 5$

When $x = 0$,
$g(0) = 2(0 + 3)^2 - 5 = 2 \times 9 - 5 = 18 - 5 = 13$

16 B

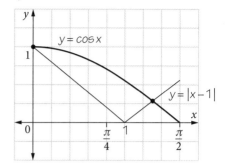

17 C

$y = (-x)^3 + (-x)$ reflection in the y-axis
$= -x^3 - x$
$= -(x - 1)^3 - (x - 1)$ translated 1 unit right

18 C

$y = 2(-x)^3 - 11(-x)^2 - 5(-x) + 8$
$= 2 \times -x^3 - 11x^2 + 5x + 8$
$= -2x^3 - 11x^2 + 5x + 8$

19 B

$y = a \times f(x)$ vertical dilation of a units
$= af(x - (-b))$ horizontal translation
$= af(x + b)$ of b units left
$= -af(x + b)$ reflection in x-axis

20 B

$e^{-x}(1 + x) = 1 - x^2$ intersect at 2 points, $(-1, 0)$ and $(0, 1)$, as shown on the diagram.

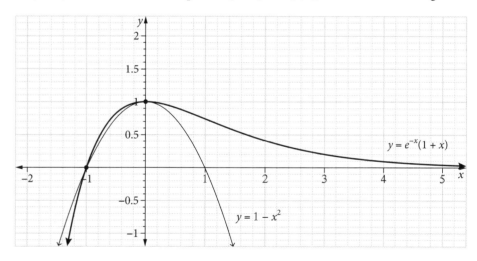

Practice set 2

Worked solutions

Question 1

$y = f(x) - 2$ with key points of
$(-3, -2), (-2, 1.5), (0, -2), (1, 1.8)$

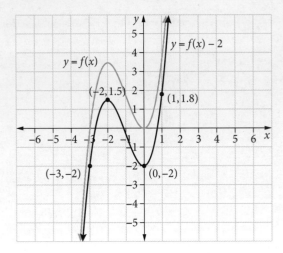

Question 2

a   Horizontal translation 3 units right, vertical dilation by scale factor of 2, vertical translation 1 unit up.

b   Horizontal translation 1 unit left, vertical translation 2 units down.

Question 3

a   Reflection in the y-axis

b
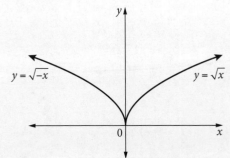

Question 4

Horizontal dilation by scale factor of $\frac{1}{3}$: $f(3x)$
$$y = 2(3x)^2 - 4 \times 3x + 3$$
$$= 2 \times 9x^2 - 12x + 3$$
$$= 18x^2 - 12x + 3$$

Reflection about y-axis: $f(-x)$
$$y = 18(-x)^2 - 12(-x) + 3$$
$$= 18x^2 + 12x + 3$$

Vertical translation 4 units up: $f(x) + 4$
$$y = 18x^2 + 12x + 3 + 4$$
$$= 18x^2 + 12x + 7$$

Question 5

a y-intercept: $x = 0$, $y = e^0 + 1 = 1 + 1 = 2$, so $(0, 2)$.

b $y = 1$

c
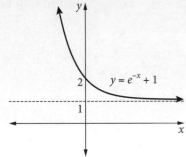

Question 6

a $y = \dfrac{1}{x + 2}$

Horizontal translation 3 units left: $f(x + 3)$
$$y = \frac{1}{(x + 3) + 2}$$
$$= \frac{1}{x + 5}$$

Vertical translation 1 unit up: $f(x) + 1$
$$y = \frac{1}{x + 5} + 1$$

b x-intercept $(y = 0)$ y-intercept $(x = 0)$

$$0 = \frac{1}{x + 5} + 1 \qquad\qquad y = \frac{1}{5} + 1$$
$$-1 = \frac{1}{x + 5} \qquad\qquad\qquad = 1\frac{1}{5}$$
$$-x - 5 = 1$$
$$-x = 6 \qquad\qquad\qquad \left(0, 1\frac{1}{5}\right)$$
$$x = -6$$

$(-6, 0)$

c
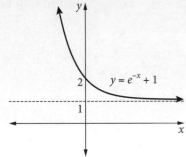

Question 7

a $f(x - 2)$ horizontal translation 2 units right
$(3 + 2, -4)$
$(5, -4)$

b $f(x + 5) + 6$ horizontal translation 5 units left
$(3, -4)$
$(3 - 5, -4)$
$(-2, -4)$
$(-2, -4 + 6)$ vertical translation 6 units up
$(-2, 2)$

Question 8

a

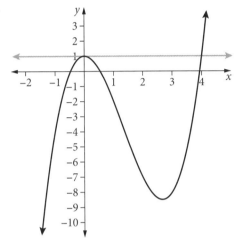

$x = 0$ and $x = 4$
(This can be checked by substitution.)

b

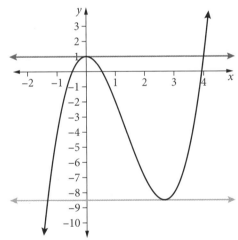

For $k = 1$ and $k = -8.5$ there are 2 solutions.
For $-8.5 < k < 1$ there are 3 solutions.
So $-8.5 \le k \le 1$.

Question 9

$y = \sqrt{8x^3 + 1}$

$y = \sqrt{8\left(\dfrac{x}{2}\right)^3 + 1}$

$y = \sqrt{x^3 + 1}$

Horizontal dilation by a scale factor of 2.

Question 10

a $f(x) = k \log_e(x + a) + c$
$f(x) = k \log_e(x + 1) + c$ is the equation because the asymptote $x = 0$ has been moved 1 unit left to $x = -1$.

So $a = 1$.

b Substitute $(0, 1)$ into the equation:

$1 = k \log_e(0 + 1) + c$
$= k \log_e(1) + c$
$= k \times 0 + c$, since $\log_e(1) = 0$

So $c = 1$, which is the y-intercept.

c Substitute $(p, 10)$ into the equation:

$$10 = k \log_e(p + 1) + 1$$
$$9 = k \log_e(p + 1)$$
$$k = \frac{9}{\log_e(p + 1)} \text{ as required.}$$

Question 11

Horizontal dilation of 4, horizontal translation 3 units right.

$x = -2 \times 4 + 3$
$= -8 + 3$
$= -5$

Vertical dilation of $\dfrac{3}{2}$, vertical translation 5 units down.

$y = \dfrac{3}{2} \times 3 - 5 = \dfrac{9}{2} - 5 = -\dfrac{1}{2}$

So $P\left(-5, -\dfrac{1}{2}\right)$.

Question 12

a Since the company is ahead in production, the function and graph needs to be translated (shifted) 1 unit left.

b $y = 400(x + 1)^3 - 2300(x + 1)^2 + 6000(x + 1) + 3000$

c i When $x = 4$,

$y = 400(4 + 1)^3 - 2300(4 + 1)^2$
$\qquad + 6000(4 + 1) + 3000$
$= 25\,500$

When $x = 5$,

$y = 400(5 + 1)^3 - 2300(5 + 1)^2$
$\qquad + 6000(5 + 1) + 3000$
$= 42\,600$

Increase in profit $= \$42\,600 - \$25\,500$
$= \$17\,100$

ii From the original graph, profit = \$20 000 when $x \approx 4.5$, that is, week 5 (any value between 4 and 5 is in week 5).

But for the transformed graph and function, this would be week 4.

A profit of \$20 000 is made in week 4.

OR by guess-and-check:

When $x = 3$,

$$y = 400(3 + 1)^3 - 2300(3 + 1)^2 + 6000(3 + 1) + 3000$$
$$= 15\,800$$

When $x = 4$, $y = \$25\,500$ from part **i**.

So profit = \$20 000 between $x = 3$ and $x = 4$, that is, in week 4 (any value between 3 and 4 is in week 4).

A profit of \$20 000 is made in week 4.

Question 13

a Horizontal translation 3 units left and vertical translation 3 units down.

b

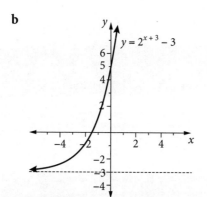

c Domain: $(-\infty, \infty)$; range: $(-3, \infty)$

Question 14

Horizontal dilation of $\frac{1}{3}$, horizontal translation 6 units right. Working backwards:

$$x = (4 - 6) \times 3 = -6$$

Vertical dilation of -1, vertical translation 10 units down. Working backwards:

$$y = (0 + 10) \times (-1) = -10$$

$$(-6, -10)$$

Question 15

$$x^2 - 6x + (-3)^2 + y^2 + 4y + 2^2 = 3 + 9 + 4$$
$$(x - 3)^2 + (y + 2)^2 = 16$$

centre $(3, -2)$, radius = 4
Reflected in the x-axis
centre: $(3, -(-2)) = (3, 2)$
equation: $(x - 3)^2 + (y - 2)^2 = 16$

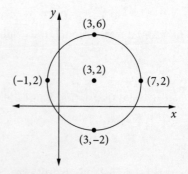

Question 16

Translated 2 units left, so $b = 2$.
Translated 1 unit down, so $c = -1$.
Substitute $(0, -7)$ into $y = k|x + 2| - 1$
$$-7 = k|0 + 2| - 1$$
$$-6 = 2k$$
$$k = -3$$

Question 17

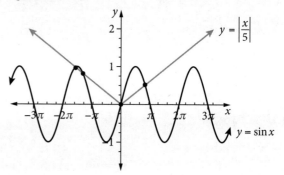

There are 4 solutions.

Question 18

a Asymptotes: $y = 1$, $x = 3$

b $f(x + 1) = \dfrac{x + 1}{x - 2}$

The graph of $f(x)$ translated 1 unit left.

When $x = 0$, $y = \dfrac{1}{-2} = -\dfrac{1}{2}$.

When $y = 0$, $0 = \dfrac{x + 1}{x - 2}$.
$$x + 1 = 0$$
$$x = -1$$

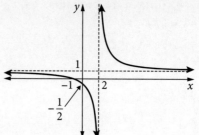

c x-intercept $(-1, 0)$, y-intercept $\left(0, -\dfrac{1}{2}\right)$

Question 19

a

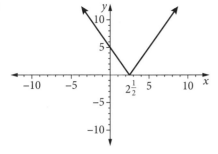

b $y = mx - 1$ has 0, 1 or 2 solutions depending on the value of m, the gradient of the line.

Note that $y = 2x - 1$ is parallel to $y = |2x - 5|$ where $x > 2.5$.

$y = -2x - 1$ is parallel to $y = |2x - 5|$ where $x < 2.5$.

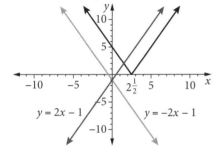

So when $-2 \leq m \leq 0$, there is no solution.

When $0 \leq m < 2$, then there are 2 or no solutions.

So to have only 1 solution, $m < -2$ or $m \geq 2$.

Question 20

a $f(x) = x^2 - 4, g(x) = 3x$

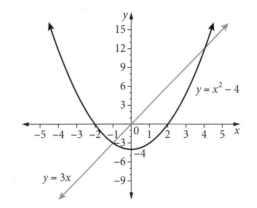

b $(-1, -3)$ and $(4, 12)$

c $x < -1$ and $x > 4$

HSC exam topic grid (2011–2020)

This table shows the coverage of this topic in past HSC exams by question number. The past exams can be downloaded from the NESA website (www.educationstandards.nsw.edu.au) by selecting 'Year 11 – Year 12', 'HSC exam papers'. NESA marking feedback and guidelines can also be found there.

Before 2020, 'Mathematics Advanced' was called 'Mathematics'. For these exams, select 'Year 11 – Year 12', 'Resources archive', 'HSC exam papers archive'.

	Graphing functions	Transformations of functions (introduced 2020)	Solving equations and inequalities graphically
2011			
2012			14(a)(iv)
2013	3, 16(b)		15(c)
2014		2	
2015	2, 12(d), 13(b)(i), 16(a)(i)–(ii)		8
2016	4, 11(a)	3	
2017	1, 11(h)	11(f)	
2018	3		
2019	12(a)		13(e)
2020 new course	**1, 2**	2, 5, **24**	11

Questions in **bold** can be found in this chapter.

CHAPTER 2
TRIGONOMETRIC FUNCTIONS

MA-T3 Trigonometric functions and graphs   29

9780170459228

TRIGONOMETRIC FUNCTIONS

Transformations of trigonometric functions

- Graphing $y = kf(a(x + b)) + c$, where $f(x) = \sin x$, $\cos x$ or $\tan x$
- Dilations
- Reflections in the x- and y-axes
- Translations
- Domain and range
- Amplitude and centre
- Period
- Phase

Trigonometric equations

- Graphical solution, including counting number of solutions
- Algebraic solution

Applications of trigonometric functions

- Practical problems involving periodic phenomena, such as tides and electric currents.

Glossary

amplitude
The height from the horizontal centre of the graph of a sine or cosine function to the peak or trough of the graph. It is half the distance between the maximum and minimum values. For $y = k \sin ax$ and $y = k \cos ax$, the amplitude is k.

centre
The mean value of a sine or cosine function that is the horizontal centre of its graph, and equidistant from the maximum and minimum values. For $y = k \sin ax$ and $y = k \cos ax$, the centre is 0. For $y = k \sin ax + c$ and $y = k \cos ax + c$, the centre is c.

period
The length of one cycle of a trigonometric function; the smallest interval on the x-axis before the function repeats itself. For $y = k \sin ax$ and $y = k \cos ax$, the period is $\dfrac{2\pi}{a}$.

https://get.ga/a-hsc-maths-advanced

A+ DIGITAL FLASHCARDS
Revise this topic's key terms and concepts by scanning the QR code or typing the URL into your browser.

phase
A horizontal shift (translation) in a transformed trigonometric function. For $y = k \sin [a(x + b)]$ and $y = k \cos [a(x + b)]$, the graphs of $y = k \sin ax$ and $y = k \cos ax$, respectively, are shifted b units to the left. The phase or phase shift is b.

translation
A transformation of a function that results in its graph being horizontally and/or vertically shifted. Its size and shape remain the same. For $y = k \sin [a(x + b)] + c$ and $y = k \cos [a(x + b)] + c$, the graphs of $y = k \sin ax$ and $y = k \cos ax$, respectively, are translated horizontally b units to the left and translated vertically c units up.

Topic summary

Trigonometric functions and graphs (MA-T3)

Vertical dilations of trigonometric functions

A vertical dilation of $y = f(x)$ is $y = kf(x)$ with scale factor k.

Example 1

The graph of $y = 3\sin x$ is a vertical stretch of $y = \sin x$.

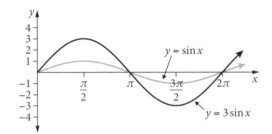

> **Hint**
> The graph is stretched vertically by a factor of k.

$y = \sin x$	$y = 3\sin x$
$(0, 0)$	$y = 3 \times 0 = 0$ $(0, 0)$
$\left(\dfrac{\pi}{2}, 1\right)$	$y = 3 \times \sin\dfrac{\pi}{2}$ $= 3$ $\left(\dfrac{\pi}{2}, 3\right)$
$(\pi, 0)$	$y = \sin\pi$ $= 0$ $(\pi, 0)$
$\left(\dfrac{3\pi}{2}, -1\right)$	$y = 3 \times \sin\dfrac{3\pi}{2}$ $= -3$ $\left(\dfrac{3\pi}{2}, -3\right)$

The range of $y = \sin x$ is $[-1, 1]$.
The range of $y = 3\sin x$ becomes $[-3, 3]$.
$k = 3$ is also the **amplitude** of $y = 3\sin x$.

Example 2

The graph of $y = -\sin x$ is $y = \sin x$ reflected in the x-axis.

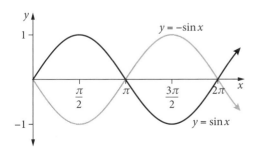

Amplitude as a vertical dilation

$y = k\sin x$ and $y = k\cos x$ have amplitude k (a vertical dilation with scale factor k).

- If $k > 1$, the function is stretched.
- If $0 < k < 1$, the function is compressed.
- If $k = -1$, the function is reflected in the x-axis.

Horizontal dilations of trigonometric functions

A horizontal dilation of $y = f(x)$ is $y = f(ax)$ with scale factor $\dfrac{1}{a}$.

Example 3

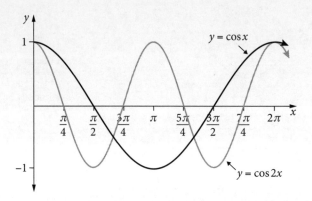

> **Hint**
>
> The graph is compressed horizontally.
>
> Period $= \dfrac{2\pi}{a}$
>
> For $[0, 2\pi]$, $y = \cos 2x$ repeats twice.

The graph of $y = \cos 2x$ is a horizontal compression of $y = \cos x$.

$y = \cos x$	$y = \cos 2x$
$(0, 1)$	$y = \cos 0 = 1$ $(0, 0)$
$\left(\dfrac{\pi}{2}, 0\right)$	$y = \cos\left(2 \times \dfrac{\pi}{2}\right)$ $= -1$ $\left(\dfrac{\pi}{2}, -1\right)$
$(\pi, -1)$	$y = \cos(2\pi)$ $= 1$ $(\pi, 1)$
$\left(\dfrac{3\pi}{2}, 0\right)$	$y = \cos\left(2 \times \dfrac{3\pi}{2}\right)$ $= -1$ $\left(\dfrac{3\pi}{2}, -1\right)$

The range of $y = \cos 2x$ does not change, $[-1, 1]$.

The value $a = 2$ is used to calculate the period of a trigonometric function.

Period $= \dfrac{2\pi}{a} = \dfrac{2\pi}{2} = \pi$.

Example 4

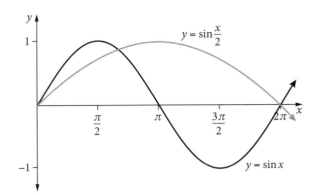

> **Hint**
>
> The graph is stretched horizontally.
>
> The period $= \dfrac{2\pi}{a}$.
>
> For $[0, 2\pi]$, $y = \sin 2x$ shows only half of the graph of $y = \sin x$.

The graph of $y = \sin \dfrac{x}{2}$ is a horizontal stretch of $y = \sin x$, maintaining the range $[-1, 1]$.

$y = \sin x$	$y = \sin\left(\dfrac{x}{2}\right)$
$x = 0, y = 0$ $(0, 0)$	$y = \sin 0 = 0$ $(0, 0)$
$\left(\dfrac{\pi}{2}, 1\right)$	$y = \sin\left(\dfrac{\frac{\pi}{2}}{2}\right)$ $= \dfrac{1}{\sqrt{2}}$ $\left(\dfrac{\pi}{2}, \dfrac{1}{\sqrt{2}}\right)$
$(\pi, 0)$	$y = \sin \dfrac{\pi}{2}$ $= 1$ $(\pi, 1)$
$\left(\dfrac{3\pi}{2}, -1\right)$	$x = \dfrac{3\pi}{2}$ $y = \sin\left(\dfrac{\frac{3\pi}{2}}{2}\right)$ $= \dfrac{1}{\sqrt{2}}$ $\left(\dfrac{3\pi}{2}, \dfrac{1}{\sqrt{2}}\right)$

The range does not change, $[-1, 1]$.

Period $= \dfrac{2\pi}{a} = \dfrac{2\pi}{\frac{1}{2}} = 4\pi$.

Example 5

The graph of $y = \cos(-x)$ is $y = \cos x$ reflected in the y-axis.

This is because every value of x changes sign.

However, note that the reflection of $y = \cos x$ in the y-axis is also itself, confirming the trigonometry property $\cos(-x) = \cos x$ (an even function).

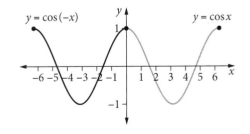

$a = -1$, the function is $y = \cos x$ reflected in the y-axis.

Period as a horizontal dilation

$$y = \sin ax \text{ has period } \frac{2\pi}{a}$$

$$y = \cos ax \text{ has period } \frac{2\pi}{a}$$

$$y = \tan ax \text{ has period } \frac{\pi}{a}$$

- If $a > 1$, the function is compressed horizontally.

- If $0 < a < 1$, the function is stretched horizontally.

- If $a = -1$, the function is reflected in the y-axis.

Vertical translations of trigonometric functions

A vertical translation of $y = f(x)$ is $y = f(x) + c$.

___

Example 6

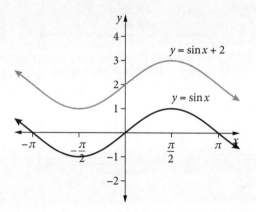

> **Hint**
> The graph $y = \sin x + 2$ shifts $y = \sin x$ vertically up 2 units.

The graph of $y = \sin x + 2$ is a vertical shift of 2 units up of $y = \sin x$. The period remains the same.

$y = \sin x$	$y = \sin x + 2$
$(0, 0)$	$y = \sin 0 + 2$ $= 2$ $(0, 2)$
$\left(\dfrac{\pi}{2}, 1\right)$	$y = \sin\left(\dfrac{\pi}{2}\right) + 2$ $= 3$ $\left(\dfrac{\pi}{2}, 3\right)$
$(\pi, 0)$	$y = \sin \pi + 2$ $= 2$ $(\pi, 2)$
$\left(\dfrac{3\pi}{2}, -1\right)$	$y = \sin\left(\dfrac{3\pi}{2}\right) + 2$ $= 1$ $\left(\dfrac{3\pi}{2}, 1\right)$

Period $= \dfrac{2\pi}{a} = \dfrac{2\pi}{1} = 2\pi$

The range of $y = \sin x$ is $[-1, 1]$.

The range of $y = \sin x + 2$ becomes $[1, 3]$. The centre, c, is 2.

Centre as a vertical translation

The centre of $y = \sin x + c$ and $y = \cos x + c$ is c.

- If $c > 0$, the centre is translated upwards.

- If $c < 0$, the centre is translated downwards.

Horizontal translations of trigonometric functions

A horizontal translation of $y = f(x)$ is $y = f(x + b)$.

Example 7

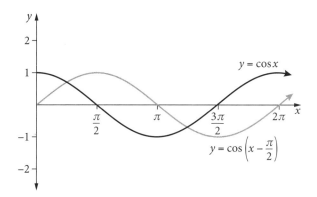

> **Hint**
>
> The graph $y = \cos\left(x - \dfrac{\pi}{2}\right)$ shows $b = \left(-\dfrac{\pi}{2}\right)$. As $b < 0$, it represents a horizontal shift of $\dfrac{\pi}{2}$ units right.

The graph of $y = \cos\left(x - \dfrac{\pi}{2}\right)$ is a horizontal shift of $\dfrac{\pi}{2}$ units right of $y = \cos x$.

The period and range do not change.

$y = \cos x$	$y = \cos\left(x - \dfrac{\pi}{2}\right)$
$(0, 1)$	$\begin{aligned} y &= \cos\left(-\dfrac{\pi}{2}\right) \\ &= 0 \\ (0, &0) \end{aligned}$
$\left(\dfrac{\pi}{2}, 0\right)$	$\begin{aligned} y &= \cos\left(\dfrac{\pi}{2} - \dfrac{\pi}{2}\right) \\ &= 1 \\ \left(\dfrac{\pi}{2}, &1\right) \end{aligned}$
$(\pi, -1)$	$\begin{aligned} y &= \cos\left(\pi - \dfrac{\pi}{2}\right) \\ &= 0 \\ (\pi, &0) \end{aligned}$
$\left(\dfrac{3\pi}{2}, 0\right)$	$\begin{aligned} y &= \cos\left(\dfrac{3\pi}{2} - \dfrac{\pi}{2}\right) \\ &= -1 \\ \left(\dfrac{3\pi}{2}, &-1\right) \end{aligned}$

Period $= \dfrac{2\pi}{a} = \dfrac{2\pi}{1} = 2\pi$.

The range is still $[-1, 1]$.

The phase, b, is $-\dfrac{\pi}{2}$.

Example 8

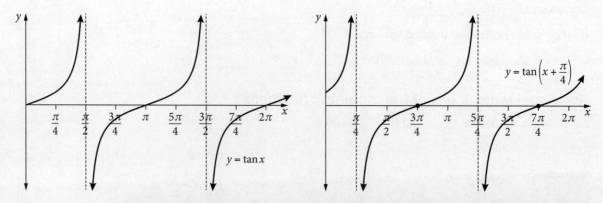

> **Hint**
>
> The graph $y = \tan\left(x + \dfrac{\pi}{4}\right)$ shows $b = \dfrac{\pi}{4}$. As $b > 0$, it represents a horizontal shift of $\dfrac{\pi}{4}$ units left.
>
> Note: The amplitude is not a feature of graphs transformed from $y = \tan x$.

The sketch of $y = \tan\left(x + \dfrac{\pi}{4}\right)$ is a horizontal shift $\dfrac{\pi}{4}$ units left of $y = \tan x$.

$y = \tan x$	$y = \tan\left(x + \dfrac{\pi}{4}\right)$
$(0, 0)$	$y = \tan\left(\dfrac{\pi}{4}\right) = 1$ $(0, 1)$
$\left(\dfrac{\pi}{4}, 1\right)$	$y = \tan\left(\dfrac{\pi}{4} + \dfrac{\pi}{4}\right) = \tan\dfrac{\pi}{2}$, which is undefined. Vertical asymptote at $x = \dfrac{\pi}{4}$.
$\left(\dfrac{\pi}{2}, 1\right)$	$y = \tan\left(\dfrac{\pi}{2} + \dfrac{\pi}{4}\right) = -1$ $\left(\dfrac{\pi}{2}, -1\right)$
$\left(\dfrac{3\pi}{4}, -1\right)$	$y = \tan\left(\dfrac{3\pi}{4} + \dfrac{\pi}{4}\right) = 0$
$(\pi, 0)$	$y = \tan\left(\pi + \dfrac{\pi}{4}\right) = 1$ $(\pi, 1)$
$\left(\dfrac{5\pi}{4}, 1\right)$	$y = \tan\left(\dfrac{5\pi}{4} + \dfrac{\pi}{4}\right) = \tan\dfrac{3\pi}{2}$, which is undefined. Vertical asymptote at $x = \dfrac{5\pi}{4}$.
$x = \dfrac{3\pi}{2}$ $y = $ undefined Vertical asymptote at $x = \dfrac{3\pi}{2}$.	$x = \dfrac{3\pi}{2}$ $y = \tan\left(\dfrac{3\pi}{2} + \dfrac{\pi}{4}\right) = -1$ $\left(\dfrac{3\pi}{2}, -1\right)$

Period $= \dfrac{\pi}{a} = \dfrac{\pi}{1} = \pi$.

The phase, b, is $\dfrac{\pi}{4}$.

Phase as a horizontal translation

The phase of $y = \sin(x + b)$, $y = \cos(x + b)$ and $y = \tan(x + b)$ is b.

- If $b > 0$, the phase shift is to the left.
- If $b < 0$, the phase shift is to the right.

Combined transformations of trigonometric functions

Function	Amplitude	Period	Phase	Centre
$y = k\sin[a(x + b)] + c$	k	$\dfrac{2\pi}{a}$	b Shift left if $b > 0$ Shift right if $b < 0$	c Shift up if $c > 0$ Shift down if $c < 0$
$y = k\cos[a(x + b)] + c$	k	$\dfrac{2\pi}{a}$		
$y = k\tan[a(x + b)] + c$	No amplitude	$\dfrac{\pi}{a}$		

Example 9

Sketch $y = 3\cos\dfrac{x}{2} - 1$ for $[0, 2\pi]$.

Solution

Amplitude and vertical dilation

Amplitude: $k = 3$

Period and horizontal dilation

$a = \dfrac{1}{2}$

$\text{Period} = \dfrac{2\pi}{\frac{1}{2}} = 4\pi$

Centre and vertical translation

Centre $= -1$

Minimum: $-1 - 3 = -4$

Maximum: $-1 + 3 = 2$

Range: $[-4, 2]$

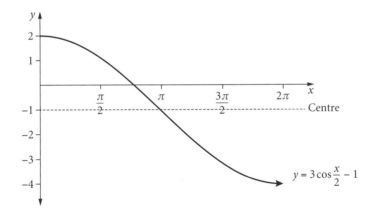

Example 10

Sketch $y = 4\sin\left(x - \dfrac{\pi}{3}\right) + 1$ for $[0, 2\pi]$.

Solution

Period	**Phase shift**	**Amplitude**	**Centre**

$a = 1$

Period $= \dfrac{2\pi}{1} = 2\pi$

$b = -\dfrac{\pi}{3}$, shift right

Amplitude: $k = 4$

Centre $= 1$

Minimum: $1 - 4 = -3$

Maximum: $1 + 4 = 5$

Range: $[-3, 5]$

Minimum point for $y = \sin x$: $y = \sin\dfrac{3\pi}{2} = -1$

So minimum point for $y = 4\sin\left(x - \dfrac{\pi}{3}\right) + 1$, when $x = \dfrac{3\pi}{2} + \dfrac{\pi}{3} = \dfrac{11\pi}{6}$ (after phase shift): $x = \dfrac{11\pi}{6}$, $y = -3$

Maximum point for $y = \sin x$: $y = \sin\dfrac{\pi}{2} = 1$

So maximum point for $y = 4\sin\left(x - \dfrac{\pi}{3}\right) + 1$, when $x = \dfrac{\pi}{2} + \dfrac{\pi}{3} = \dfrac{5\pi}{6}$: $x = \dfrac{5\pi}{6}$, $y = 5$

y-intercept: when $x = 0$,

$y = 4\sin\left(0 - \dfrac{\pi}{3}\right) + 1$

≈ -2.5

$(0, -2.5)$

Also, when $x = 2\pi$, $y \approx -2.5$.

$(2\pi, -2.5)$

Translations from $y = \sin x$

$(0, 0)$	becomes	$\left(0 + \dfrac{\pi}{3}, 0 + 1\right) = \left(\dfrac{\pi}{3}, 1\right)$
$(\pi, 0)$	becomes	$\left(\pi + \dfrac{\pi}{3}, 0 + 1\right) = \left(\dfrac{4\pi}{3}, 1\right)$

x	0	$\dfrac{\pi}{3}$	$\dfrac{5\pi}{6}$	$\dfrac{4\pi}{3}$	$\dfrac{11\pi}{6}$	2π
y	-2.5	1	5	1	-3	-2.5

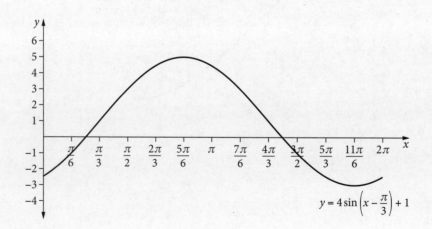

$y = 4\sin\left(x - \dfrac{\pi}{3}\right) + 1$

Trigonometric equations

We can solve trigonometric equations graphically for a given domain.

Example 11

The graph of the trigonometric function

$y = 2\sin\left[3\left(x - \dfrac{\pi}{3}\right)\right] + 1$ is shown for $[0, 2\pi]$.

Find the number of solutions to the trigonometric

equation $2\sin\left[3\left(x - \dfrac{\pi}{3}\right)\right] + 1 = 0$ for $[0, 2\pi]$.

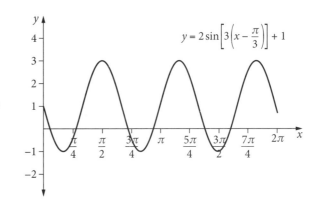

Solution

Locate the point(s) where the graph of the trigonometric equation cuts the x-axis (the x-intercepts).

There are 6 x-intercepts for the given domain. These are the solutions for $[0, 2\pi]$.

There are 6 solutions.

Example 12

Sketch the graphs of $y = \dfrac{x}{4} - 1$ and $y = 3\cos x - 2$ for $[0, 2\pi]$.

Hence, determine how many solutions there are for the equation $\dfrac{x}{4} - 1 = 3\cos x - 2$ for $[0, 2\pi]$.

Solution

$y = 3\cos x - 2$ has amplitude $k = 3$ and centre $c = -2$.

Minimum: $-2 - 3 = -5$
Maximum: $-2 + 3 = 1$

Period $= 2\pi$

$y = \dfrac{x}{4} - 1$ is a straight line with x-intercept 4 and

y-intercept -1.

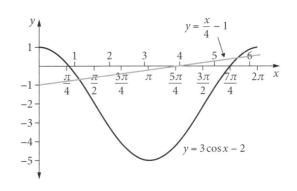

Locate where the graphs cross each other.

These are the solutions for $[0, 2\pi]$.

There are 2 solutions.

Example 13

Solve each equation for $[0, 2\pi]$:

a $\sin 2x = 1$

b $\cos\left(2x - \dfrac{\pi}{4}\right) = \dfrac{1}{\sqrt{2}}$

Solution

a $\sin 2x = 1$

$$2x = \frac{\pi}{2}, 2\pi + \frac{\pi}{2} \qquad \text{2 revolutions}$$

$$= \frac{\pi}{2}, \frac{5\pi}{2}$$

$$x = \frac{\pi}{4}, \frac{5\pi}{4}$$

b $\cos\left(2x - \dfrac{\pi}{4}\right) = \dfrac{1}{\sqrt{2}}$ positive in the 1st and 4th quadrants

$$2x - \frac{\pi}{4} = \frac{\pi}{4}, 2\pi - \frac{\pi}{4}, 2\pi + \frac{\pi}{4}, 4\pi - \frac{\pi}{4} \qquad \text{2 revolutions}$$

$$= \frac{\pi}{4}, \frac{7\pi}{4}, \frac{9\pi}{4}, \frac{15\pi}{4}$$

$$2x = \frac{\pi}{4} + \frac{\pi}{4}, \frac{7\pi}{4} + \frac{\pi}{4}, \frac{9\pi}{4} + \frac{\pi}{4}, \frac{15\pi}{4} + \frac{\pi}{4}$$

$$= \frac{\pi}{2}, 2\pi, \frac{5\pi}{2}, 4\pi$$

$$x = \frac{\pi}{4}, \pi, \frac{5\pi}{4}, 2\pi$$

Example 14

Solve each equation for $[0, 2\pi]$:

a $\sqrt{3}\tan 3x = 1$

b $\sin\left(x - \dfrac{\pi}{2}\right) = \dfrac{\sqrt{3}}{2} \qquad 0 \le x \le 2\pi$

Solution

a $\tan 3x = \dfrac{1}{\sqrt{3}}$ positive in 1st and 3rd quadrants

$$3x = \frac{\pi}{6}, \pi + \frac{\pi}{6}, 2\pi + \frac{\pi}{6}, 3\pi + \frac{\pi}{6}, 4\pi + \frac{\pi}{6}, 5\pi + \frac{\pi}{6} \qquad \text{3 revolutions}$$

$$= \frac{\pi}{6}, \frac{7\pi}{6}, \frac{13\pi}{6}, \frac{19\pi}{6}, \frac{25\pi}{6}, \frac{31\pi}{6}$$

$$x = \frac{\pi}{18}, \frac{7\pi}{18}, \frac{13\pi}{18}, \frac{19\pi}{18}, \frac{25\pi}{18}, \frac{31\pi}{18}$$

b $\sin\left(x - \dfrac{\pi}{2}\right) = \dfrac{\sqrt{3}}{2} \qquad 0 \le x \le 2\pi$

$$-\frac{\pi}{2} \le x - \frac{\pi}{2} \le \frac{3\pi}{2}$$

The new domain is $\left[\dfrac{-\pi}{2}, \dfrac{3\pi}{2}\right]$.   1 revolution starting at $-\dfrac{\pi}{2}$

$$\sin\left(x - \frac{\pi}{2}\right) = \frac{\sqrt{3}}{2}$$ $\sin x > 0$ in 1st and 2nd quadrants

$$x - \frac{\pi}{2} = \frac{\pi}{3}, \pi - \frac{\pi}{3}$$

$$= \frac{\pi}{3}, \frac{2\pi}{3}$$

$$x = \frac{\pi}{3} + \frac{\pi}{2}, \frac{2\pi}{3} + \frac{\pi}{2}$$

$$= \frac{5\pi}{6}, \frac{7\pi}{6}$$

Practice set 1

Multiple-choice questions

Solutions start on page 48.

Question 1

What is the amplitude and period of $y = \cos \pi x$?

A amplitude = 1, period = 2

B amplitude = π, period = 2

C amplitude = 1, period = 2π

D amplitude = π, period = 2π

Question 2

The function $f(x) = \sin x$ is transformed to $g(x) = \sin\left(x - \frac{\pi}{2}\right) - 1$ by a horizontal translation of $\frac{\pi}{2}$ units, followed by a vertical translation of 1 unit.

Which row of the table shows the directions of the translations?

	Horizontal translation of $\frac{\pi}{2}$	Vertical translation of 1 unit
A	Left	Up
B	Right	Up
C	Left	Down
D	Right	Down

Question 3

What is the phase of the function $y = \frac{1}{3}\sin\left(2x + \frac{\pi}{4}\right)$?

A $\frac{1}{3}$

B $\frac{\pi}{8}$

C $\frac{1}{2}$

D $\frac{\pi}{4}$

Question 4

Which one of the following statements is true for $y = \cos\frac{x}{3}$ for $0 \le x \le 6\pi$?

A The amplitude is 3.

B The range is $\left[-\frac{1}{3}, \frac{1}{3}\right]$.

C The period is $\frac{2\pi}{3}$.

D It has 2 x-intercepts.

9780170459228

Question 5   ©NESA   2020 HSC EXAM, QUESTION 6   ◉ ○ ○

Which interval gives the range of the function $y = 5 + 2\cos 3x$?

A $[2, 8]$ **B** $[3, 7]$

C $[4, 6]$ **D** $[5, 9]$

Question 6   ◉ ○ ○

What is the period of the function $f(x) = \tan 4x$?

A $\dfrac{\pi}{4}$ **B** $\dfrac{\pi}{2}$

C 4π **D** 8π

Question 7   ◉ ○ ○

The equation of the transformation of $y = \sin x$ has a phase of $\dfrac{\pi}{4}$ units left and a vertical dilation with scale factor 5.

What is the transformed function?

A $y = 5\sin\left(x - \dfrac{\pi}{4}\right)$ **B** $y = \sin\left(x - \dfrac{\pi}{4}\right) + 5$

C $y = 5\sin\left(x + \dfrac{\pi}{4}\right)$ **D** $y = \sin\left(x + \dfrac{\pi}{4}\right) + 5$

Question 8   ◉ ○ ○

The diagram shows part of the graph of $y = a\cos(bx) - 1$.

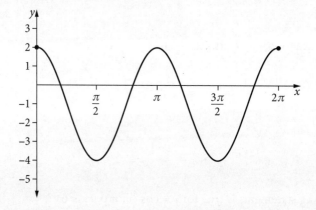

What are the values of a and b?

A $a = 6, b = 2$ **B** $a = 6, b = \dfrac{1}{2}$

C $a = 3, b = 2$ **D** $a = 3, b = \dfrac{1}{2}$

Question 9   ◉ ○ ○

Which one of the following statements is true if $f(x) = \cos 2x$ is transformed to $g(x) = \cos 2\left(x + \dfrac{\pi}{2}\right)$?

A They represent the same graph.   **B** $g(x)$ is a reflection of $f(x)$ in the x-axis.

C $g(x)$ is a vertical translation of π.   **D** The period is halved.

Question 10

Which diagram shows the graph of $y = \sin\left(2x + \dfrac{\pi}{4}\right)$?

A

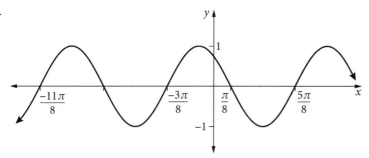

B

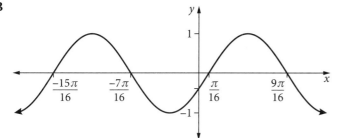

C

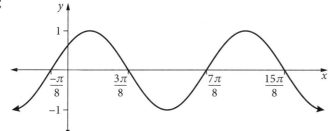

D

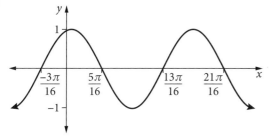

Question 11

What is the solution of $\cos\dfrac{x}{2} = 1$ for $0 \le x \le 2\pi$?

A $x = 0$

B $x = 2\pi$

C $x = 0, 2\pi$

D $x = 0, 4\pi$

Questions 12 and 13 relate to the information and diagram given below.

The depth of water in a harbour is given by $y = 10 + 4\sin\dfrac{t}{2}$, where t is the number of hours after midnight.

The function is graphed below for one 24-hour period.

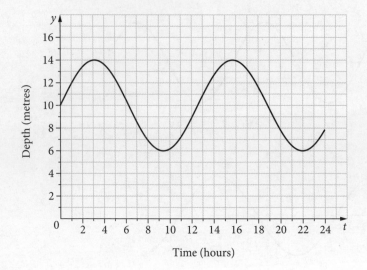

Time (hours)

Question 12 ⬤⬤⬛

What is the exact time, in hours, when the water is first at its maximum depth during the 24-hour period?

A $\dfrac{\pi}{2}$

B π

C 9.5

D 10

Question 13 ⬤⬤⬛

Ships can enter the harbour when there is a water depth of 7 metres or more in the harbour. At all other times, the harbour gates are closed.

For how many hours in a 24-hour period are the harbour gates closed?

A 6

B 7

C 18

D 19

Question 14 ⬤⬤⬛

The curve $y = k\sin ax + b$ (where $0 \le x \le \pi$) has a maximum value at 1, amplitude 2 and a period of $\dfrac{5\pi}{6}$.

What are the values of k, a and b?

A $k = 2, a = \dfrac{12}{5}, b = -1$

B $k = \dfrac{1}{2}, a = \dfrac{5}{12}, b = 1$

C $k = 2, a = \dfrac{12}{5}, b = 3$

D $k = \dfrac{1}{2}, a = \dfrac{5}{12}, b = 3$

Question 15

What are the period and range of the function $f(x) = 4\sin\left(\dfrac{3x}{5}\right) - 3$?

	Period	Range
A	$\dfrac{10\pi}{3}$	$[-4, 4]$
B	$\dfrac{10\pi}{3}$	$[-7, 1]$
C	$\dfrac{6\pi}{5}$	$[-1, 7]$
D	$\dfrac{6\pi}{5}$	$[-4, 4]$

Question 16

How many solutions does the equation $\left|\dfrac{1}{2}\sin x\right| = \dfrac{1}{4}$ have for $0 \le x \le 2\pi$?

A 2   **B** 4

C 6   **D** 8

Question 17

The UV index is a measure of the level of solar UV developed by the World Health Organization to raise awareness about the risk of overexposure to UV in sunlight. The graph below shows the UV index measured hourly from 6 am (represented by $t = 0$) on a particular January day in Sydney.

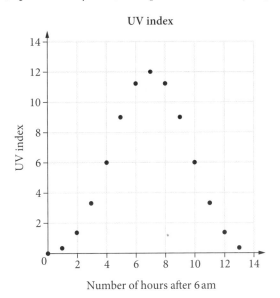

UV index

Number of hours after 6 am

Data obtained from: arpansa.gov.au

Which equation best represents the graph?

A $y = 6 + 6\sin\dfrac{\pi t}{14}$ **B** $y = 6 + 6\sin\dfrac{\pi t}{7}$

C $y = 6 - 6\cos\dfrac{\pi t}{14}$ **D** $y = 6 - 6\cos\dfrac{\pi t}{7}$

Question 18 ●●●

What is the solution to $\tan 2x = -\dfrac{1}{\sqrt{3}}$ for $0 < x < \pi$?

A $x = \dfrac{3\pi}{8}, \dfrac{7\pi}{8}, \dfrac{11\pi}{8}, \dfrac{15\pi}{8}$

B $x = \dfrac{3\pi}{8}, \dfrac{7\pi}{8}$

C $x = \dfrac{5\pi}{12}, \dfrac{11\pi}{12}, \dfrac{17\pi}{12}, \dfrac{23\pi}{12}$

D $x = \dfrac{5\pi}{12}, \dfrac{11\pi}{12}$

Question 19 ●●○

The face of an analogue clock has a diameter of 40 cm and its minute hand is 15 cm long, as shown in the diagram below.

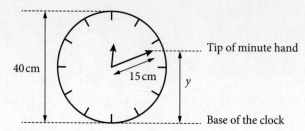

At 12 pm, both hands of the clock point vertically up and the tip of the minute hand is at its maximum height above the base of the clock.

Which equation gives the height, y centimetres, of the tip of the minute hand above the base of the clock, t minutes after 12 pm?

A $y = 20 + 15\sin\dfrac{\pi t}{30}$

B $y = 40 - 15\sin\dfrac{\pi t}{60}$

C $y = 20 + 15\cos\dfrac{\pi t}{30}$

D $y = 40 - 15\cos\dfrac{\pi t}{60}$

Question 20 ●●●

The sum of the solutions of $\tan\left(2x + \dfrac{\pi}{3}\right) = -1$ over the interval $[-\pi, m]$ is $-\dfrac{\pi}{6}$.

Which of the following is a possible value of m?

A $\dfrac{5\pi}{24}$

B $\dfrac{5\pi}{6}$

C $\dfrac{4\pi}{3}$

D $\dfrac{3\pi}{2}$

Practice set 2

Short-answer questions

Solutions start on page 51.

Question 1 (1 mark) ●○○

Sketch the graph of $y = 3\cos 2x$ for the domain $[0, 2\pi]$.   1 mark

Question 2 (3 marks) ●○○

The diagram shows part of the graph of $f(x) = c + k\sin ax$.
The period is 4π, the minimum value is 3 and the
maximum value is 11.

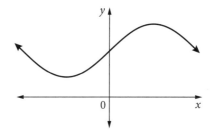

Find the value of c, k and a in the above function.   3 marks

Question 3 (2 marks) ●○○

Determine the range of the function $y = 4\sin\dfrac{2\pi x}{3} + 1$.   2 marks

Question 4 (5 marks) ●○○

Find the equation of the transformed function when $y = \cos x$ is vertically dilated by a scale   5 marks
factor of 4, horizontally dilated by a scale factor of 2, vertically translated 3 units down and

horizontally translated $\dfrac{\pi}{6}$ units to the left.

Question 5 (4 marks) ●●○

Consider the equation $y = 9 - 4\sin(2x - \pi)$.

a Determine the amplitude, period, phase and vertical translation.   2 marks

b Hence, sketch its graph for $[0, 2\pi]$.   2 marks

Question 6 (2 marks) ●●○

Solve $\sin\dfrac{\theta}{2} = \dfrac{\sqrt{3}}{2}$ for $0 \le \theta \le 2\pi$.   2 marks

Question 7 (3 marks) ●●○

a Sketch the graphs of $y = \cos x - 1$ and $y = 1 - x$ for $[0, 2\pi]$.   2 marks

b Hence, solve for x: $\cos x - 1 = 1 - x$ for $[0, 2\pi]$.   1 mark

Question 8 (2 marks) ●●○

Solve $\sin\left(x - \dfrac{\pi}{4}\right) = \dfrac{1}{2}$ for $0 \le x \le 2\pi$.   2 marks

Question 9 (2 marks) ●●○

Find the period and range of $f(x) = 3\cos\dfrac{x}{3} + \pi$. Leave your answers in terms of π.   2 marks

Question 10 (3 marks) ●●○

Solve $\tan 3\theta = -1$ for $0 \le \theta \le 2\pi$.   3 marks

Question 11 (4 marks)

The water in a creek fluctuates according to the tide. At time t hours after midnight, the height of the tide, h metres, is given by the equation

$$h = 1 + 0.5\sin\left[\frac{\pi}{3}(t + 0.2)\right].$$

a   Calculate the height of the tide at 8:00 am, correct to three decimal places.   2 marks

b   Calculate the exact time when the tide first reaches a height of 1 m in the 24-hour period.   2 marks

Question 12 (3 marks)

Azaa, Ben and Chloe were discussing the graph shown on the right.

They did not agree on the trigonometric equation for this graph. Their equations are shown below.

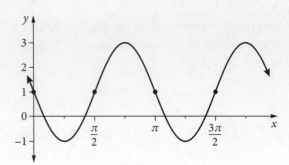

Azaa $y = 2\cos\left(2x + \frac{\pi}{2}\right) + 1$

Ben $y = -2\sin\left[2\left(x - \frac{\pi}{2}\right)\right] + 1$

Chloe $y = -2\sin 2x + 1$

Compare the equations to decide which equation(s) match the graph. Show all working to justify your answer.   3 marks

Question 13 (4 marks)

The graph of $y = \sin x$ is transformed to $y = 4\sin[3(2 - x)] + 1$.

For the new function, describe: 4 marks

• the centre, maximum and minimum values

• the horizontal and/or vertical dilation and scale factors

• reflection in either the x- and/or y-axis

• the phase shift of the curve.

A sketch and intercepts are not required, but may be drawn as an aid to answer the question.

Question 14 (3 marks)

The graph of $f(x) = 2\cos x$, for $0 \le x \le 2\pi$ is shown in the diagram on the right. The graph intersects the line $y = -1$ at points A and B.

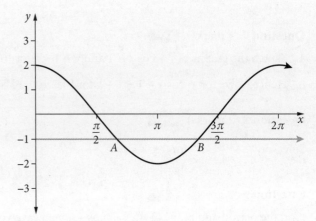

a   Find the x-coordinates of points A and B.   2 marks

b   If $g(x) = 2\cos px$, $0 \le x \le 2\pi$, where p is a positive integer, find the value of p so that $g(x) = -1$ has exactly 6 solutions.   1 mark

Question 15 (3 marks)

Solve $\sqrt{2}\cos\left[2\left(x-\frac{\pi}{8}\right)\right]=1$ for $0 \le x \le 2\pi$. 3 marks

Question 16 (8 marks)

a Find the range of $f(x) = 2\sin 3x + 4$. 2 marks

b Consider $g(x) = 5f(2x)$.

   i Find its range. 2 marks

   ii Given that $y = g(x)$ can be written in the form $g(x) = 10\sin ax + c$, find a and c. 2 marks

   iii $g(x) = 12$ has two solutions $\pi \le x \le \frac{4\pi}{3}$. Find the two values for x. 2 marks

Question 17 (3 marks)

Solve the equation $2\cos^2 x = 3\sin x$, for $0 \le x \le \pi$. Answer in terms of π. 3 marks

Question 18 (3 marks)

a Sketch $y = \tan 2x$ and $y = x - \pi$, for $-\pi \le x \le \pi$, on the same axes. 2 marks

b Using the graph in part **a**, state the number of solutions to the equation $\tan 2x + \pi = x$. 1 mark

Question 19 (4 marks)

a Sketch $f(x) = 3\sin\left(2x + \frac{\pi}{4}\right) + 2$ for $-\pi \le x \le \pi$. 3 marks

b Hence, state the number of solutions to $3\sin\left(2x + \frac{\pi}{4}\right) + 2 = -\frac{1}{2}x + 1$. 1 mark

Question 20 (5 marks) ©NESA 2020 HSC EXAM, QUESTION 31

The population of mice on an isolated island can be modelled by the function

$$m(t) = a\sin\left(\frac{\pi}{26}t\right) + b,$$

where t is the time in weeks and $0 \le t \le 52$. The population of mice reaches a maximum of 35 000 when $t = 13$ and a minimum of 5000 when $t = 39$. The graph of $m(t)$ is shown.

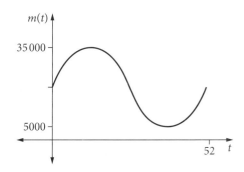

a What are the values of a and b? 2 marks

b On the same island, the population of cats can be modelled by the function

$$c(t) = -80\cos\left(\frac{\pi}{26}(t-10)\right) + 120.$$

Consider the graph of $m(t)$ and the graph of $c(t)$. Find the values of t, $0 \le t \le 52$, for which both populations are increasing. 3 marks

Practice set 1

Worked solutions

1 A

Amplitude $= 1$ Period $= \dfrac{2\pi}{\pi} = 2$

2 D

Horizontal translation $\dfrac{\pi}{2}$ units right, and vertical translation 1 unit down.

3 B

For $y = \dfrac{1}{3}\sin\left(2x + \dfrac{\pi}{4}\right)$

$\qquad = \dfrac{1}{3}\sin\left[2\left(x + \dfrac{\pi}{8}\right)\right]$

Phase $= \dfrac{\pi}{8}$

4 D

Amplitude $= 1$

Range $= [-1, 1]$

Period $= 6\pi$

$y = \cos\dfrac{x}{3}$ for $0 \le x \le 6\pi$ is dilated horizontally. It has only 2 x-intercepts.

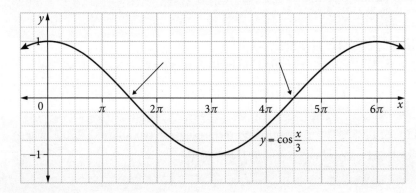

5 B

The range for $y = \cos x$ is $[-1, 1]$.

So for $y = 5 + 2\cos 3x$:

minimum value: $5 + 2 \times (-1) = 3$
maximum value: $5 + 2 \times 1 = 7$

Range is $[3, 7]$.

6 A

$$\text{Period} = \frac{\pi}{a}, \text{ where } a = 4$$

$$= \frac{\pi}{4}$$

7 C

$$y = 5\sin\left(x + \frac{\pi}{4}\right)$$

8 C

Amplitude $= 6 \div 2 = 3$

Period $= \pi$

9 B

Reflection in the x-axis.

$$y = \cos 2\left(x + \frac{\pi}{2}\right) = \cos(2x + \pi)$$

When $x = 0$, $y = \cos 2x = \cos 0 = 1$
and so $y = \cos(2x + \pi)$

$$= \cos((2 \times 0) + \pi)$$

$$= -1$$

10 C

$$y = \sin\left(2x + \frac{\pi}{4}\right)$$

Horizontal translation $\frac{\pi}{4}$ to the left

Period $= 2$

11 A

$$\cos\frac{x}{2} = 1 \text{ for } 0 \le x \le 2\pi$$

$$\frac{x}{2} = 0, 2\pi$$

$$x = 0, 4\pi$$

Since $0 \le x \le 2\pi$,
$x = 0$ only.

$$\frac{x}{2} = \frac{\pi}{3}, \frac{5\pi}{3}$$

$$x = \frac{2\pi}{3}, \frac{10\pi}{3}$$

But $x = \frac{10\pi}{3} > 2\pi$

so $x = \frac{2\pi}{3}$ only.

12 B

$$10 + 4\sin\frac{t}{2} = 14$$

$$4\sin\frac{t}{2} = 4$$

$$\sin\frac{t}{2} = 1$$

$$\frac{t}{2} = \frac{\pi}{2}$$

$$t = \pi$$

13 A

Depth is less than 7 metres from 8 am to 11 am, and 8:30 pm to 11:30 pm.

$3 + 3 = 6$ hours

14 A

Amplitude $= 2$, so $k = 2$

$$\text{Period} = \frac{2\pi}{a} = \frac{5\pi}{6}$$

$$a = 2\pi \times \frac{6}{5\pi} = \frac{12\pi}{5\pi} = \frac{12}{5}$$

Maximum value $= 1$

$$2 + b = 1$$

$$b = -1$$

15 B

$$\text{Period} = \frac{2\pi}{\left(\frac{3}{5}\right)} = \frac{10\pi}{3}$$

Range for $y = \sin x$ is $[-1, 1]$

so for $f(x) = 4\sin\left(\frac{3x}{5}\right) - 3$:

minimum value: $4 \times (-1) - 3 = -7$
maximum value: $4 \times 1 - 3 = 1$

Range is $[-7, 1]$.

16 B

There are 4 solutions.

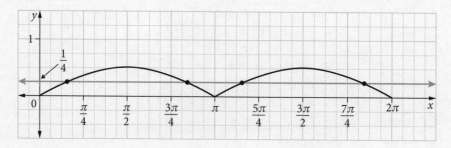

17 D

$$\text{Period} = \frac{2\pi}{n} = 14$$

$$\frac{2\pi}{14} = n$$

$$n = \frac{\pi}{7}$$

Equations B and D have $n = \frac{\pi}{7}$.

From the graph, when $t = 0$, $x = 0$.

B $x = 6 + 6\sin 0 = 6$

D $x = 6 - 6\cos 0 = 6 - 6$
$$= 0$$

18 D

$$\tan 2x = -\frac{1}{\sqrt{3}} \text{ for } 0 < x < \pi$$

$$2x = \pi - \frac{\pi}{6}, 2\pi - \frac{\pi}{6}, 2\pi + \frac{5\pi}{6}, 2\pi + \frac{11\pi}{6}$$

$$= \frac{5\pi}{6}, \frac{11\pi}{6}, \frac{17\pi}{6}, \frac{23\pi}{6}$$

$$x = \frac{5\pi}{12}, \frac{11\pi}{12}, \frac{17\pi}{12}, \frac{23\pi}{12}$$

But $0 < x < \pi$, so $x = \frac{5\pi}{12}, \frac{11\pi}{12}$ only.

19 C

Diameter = 40 cm, radius = 20 cm

Amplitude = 15 cm

$$\text{Period} = \frac{2\pi}{a} = 60$$

$$a = \frac{2\pi}{60} = \frac{\pi}{30}$$

$$y = 20 + 15\cos\left(\frac{\pi t}{30}\right)$$

20 B

$$\tan\left(2x + \frac{\pi}{3}\right) = -1$$

$$2x + \frac{\pi}{3} = \frac{3\pi}{4}, \frac{7\pi}{4}, \frac{3\pi}{4} - \pi, \frac{3\pi}{4} - 2\pi$$

$$= \frac{3\pi}{4}, \frac{7\pi}{4}, \frac{-\pi}{4}, \frac{-5\pi}{4}$$

$$2x = \frac{3\pi}{4} - \frac{\pi}{3}, \frac{7\pi}{4} - \frac{\pi}{3}, \frac{-\pi}{4} - \frac{\pi}{3}, \frac{-5\pi}{4} - \frac{\pi}{3}$$

$$= \frac{5\pi}{12}, \frac{17\pi}{12}, \frac{-7\pi}{12}, \frac{-19\pi}{12}$$

$$x = \frac{5\pi}{24}, \frac{17\pi}{24}, \frac{-7\pi}{24}, \frac{-19\pi}{24}$$

$$\frac{5\pi}{24} + \frac{17\pi}{24} + \frac{-7\pi}{24} + \frac{-19\pi}{24} = \frac{-4\pi}{24} = \frac{-\pi}{6}$$

$$m = \frac{5\pi}{6}$$

Practice set 2

Worked solutions

Question 1

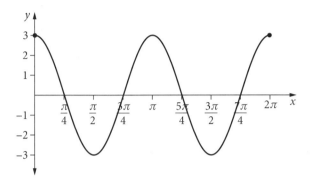

Question 2

$c = 7$

$k = 4$

Period $= 4\pi = \dfrac{2\pi}{a}$

$a = \dfrac{2\pi}{4\pi} = \dfrac{1}{2}$

Question 3

$y = \sin x$; Range: $[-1, 1]$

$y = 4\sin\left(\dfrac{2\pi x}{3}\right) + 1$; Range: $[-4 + 1, 4 + 1] = [-3, 5]$

Question 4

$y = 4\cos\left[\dfrac{1}{2}\left(x + \dfrac{\pi}{6}\right)\right] - 3$

Question 5

a　Amplitude is 4.

$y = 9 - 4\sin(2x - \pi)$ becomes

$y = 9 - 4\sin\left[2\left(x - \dfrac{\pi}{2}\right)\right]$

so period $= \dfrac{2\pi}{a} = \dfrac{2\pi}{2} = \pi$ and

phase $= -\dfrac{\pi}{2}$.

Vertical shift is 9 units up.

b

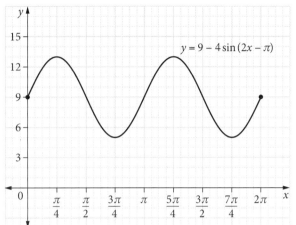

Question 6

$\sin\left(\dfrac{\theta}{2}\right) = \dfrac{\sqrt{3}}{2}$　for $0 \le \theta \le 2\pi$

where $0 \le \dfrac{\theta}{2} \le \pi$.

So $\dfrac{\theta}{2} = \dfrac{\pi}{3}, \dfrac{2\pi}{3}$

$\theta = \dfrac{2\pi}{3}, \dfrac{4\pi}{3}$

Question 7

a

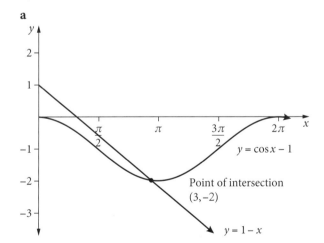

b　Diagram above shows the solution for x:

$x \approx 3$

Question 8

$$x - \frac{\pi}{4} = \frac{\pi}{6}, \frac{5\pi}{6}$$

$$x = \frac{\pi}{6} + \frac{\pi}{4}, \frac{5\pi}{6} + \frac{\pi}{4}$$

$$= \frac{5\pi}{12}, \frac{13\pi}{12}$$

Question 9

$f(x) = 3\cos\left(\frac{x}{3}\right) + \pi$ is $f(x) = k\cos(ax) + c$

where $k = 3$, $a = \frac{1}{3}$ and $c = \pi$.

Period: $\dfrac{2\pi}{a} = \dfrac{2\pi}{\left(\frac{1}{3}\right)} = 6\pi$

Minimum value $= 3(-1) + \pi = \pi - 3$
Maximum value $= 3(1) + \pi = \pi + 3$

Range: $[\pi - 3, \pi + 3]$

Question 10

$\tan 3\theta = -1$

$$3\theta = \frac{3\pi}{4}, \frac{7\pi}{4}, 2\pi + \frac{3\pi}{4}, 2\pi + \frac{7\pi}{4}, 4\pi + \frac{3\pi}{4}, 4\pi + \frac{7\pi}{4}$$

$$= \frac{3\pi}{4}, \frac{7\pi}{4}, \frac{11\pi}{4}, \frac{15\pi}{4}, \frac{19\pi}{4}, \frac{23\pi}{4}$$

$$\theta = \frac{\pi}{4}, \frac{7\pi}{12}, \frac{11\pi}{12}, \frac{5\pi}{4}, \frac{19\pi}{12}, \frac{23\pi}{12}$$

Question 11

a 8:00 am is 8 hours after midnight.

Substitute $t = 8$:

$$h = 1 + 0.5\sin\left[\frac{\pi}{3}(8 + 0.2)\right]$$

$$\approx 1.372 \text{ metres}$$

b
$$1 = 1 + 0.5\sin\left[\frac{\pi}{3}(t + 0.2)\right]$$

$$0 = 0.5\sin\left[\frac{\pi}{3}(t + 0.2)\right]$$

$$= \sin\left[\frac{\pi}{3}(t + 0.2)\right]$$

$$\frac{\pi}{3}(t + 0.2) = \pi$$

$$t + 0.2 = 3$$

$$t = 2.8\,\text{h}$$

$$= 2\,\text{h}\,48\,\text{min}$$

12:00 am + 2 h 48 min = 2:48 am

Question 12

Azaa:

$$y = 2\cos\left(2x + \frac{\pi}{2}\right) + 1\, y = 2\cos\left[2\left(x + \frac{\pi}{4}\right)\right] + 1$$

Amplitude = 2

Vertical translation = 1 unit up

Period = π

Phase: shift left $\dfrac{\pi}{4}$

These features match the graph shown.
Azaa is correct.

Ben:

$$y = -2\sin\left[2\left(x - \frac{\pi}{2}\right)\right] + 1$$

Amplitude 2, but reflected in the x-axis

Vertical translation = 1 unit up

Period = π

Phase: shift left $\dfrac{\pi}{2}$

The shift needed to be $\dfrac{\pi}{4}$ left to match the graph

shown. Ben is incorrect.

Chloe:
$$y = -2\sin 2x + 1$$

Amplitude 2, but reflected in the x-axis

Vertical translation = 1 unit up

Period = π
Phase = 0

These features match the graph shown.
Chloe is correct.

Azaa and Chloe are both correct.

Question 13

$$y = 4\sin\left[3(2 - x)\right] + 1$$
$$= 4\sin\left[-3(x - 2)\right] + 1$$
$$= k\sin\left[a(x + b)\right] + c$$

Centre = 1

Maximum value $= 1 + 4 = 5$
Minimum value $= 1 - 4 = -3$

Horizontal dilation $a = -3$ includes a reflection in

y-axis: scale factor of $\dfrac{1}{3}$

Vertical dilation: scale factor = 4

Phase shift: -2

Question 14

a $2\cos x = -1$

$\cos x = -\dfrac{1}{2}$　　　has solutions in the 2nd and 3rd quadrants

$x = \dfrac{2\pi}{3}, \dfrac{4\pi}{3}$

b $2\cos x = -1$ has 2 solutions, from part **a**.

We want $2\cos px = -1$ to have 6 solutions. This is only possible if we compress the graph of $y = 2\cos x$ horizontally by scale factor $\dfrac{1}{3}$.

So $p = 3$ for 6 solutions.

Question 15

$\sqrt{2}\cos\left[2\left(x - \dfrac{\pi}{8}\right)\right] = 1$

$\cos\left[2\left(x - \dfrac{\pi}{8}\right)\right] = \dfrac{1}{\sqrt{2}}$

cos is positive in 1st and 4th quadrants.

$2\left(x - \dfrac{\pi}{8}\right) = -\dfrac{\pi}{4}, \dfrac{\pi}{4}, \dfrac{7\pi}{4}, 2\pi + \dfrac{\pi}{4}, 2\pi + \dfrac{7\pi}{4}$

　　　　　　(2 revolutions)

$x - \dfrac{\pi}{8} = -\dfrac{\pi}{8}, \dfrac{\pi}{8}, \dfrac{7\pi}{8}, \dfrac{9\pi}{8}, \dfrac{15\pi}{8}, \dfrac{17\pi}{8}$

$x = 0, \dfrac{\pi}{4}, \pi, \dfrac{5\pi}{4}, 2\pi, \dfrac{9\pi}{4}$

But $\dfrac{9\pi}{4} > 2\pi$, so $x = 0, \dfrac{\pi}{4}, \pi, \dfrac{5\pi}{4}, 2\pi$.

Question 16

a Minimum $= 2\sin\left(3 \times \dfrac{\pi}{2}\right) + 4$

$= 2(-1) + 4$

$= 2$

Maximum $= 2\sin\left(3 \times \dfrac{\pi}{6}\right) + 4$

$= 2(1) + 4$

$= 6$

Range is $[2, 6]$.

b　i Use range from part **a**, $[2, 6]$ with vertical dilation $= 5$.

Minimum: $5 \times 2 = 10$

Maximum: $5 \times 6 = 30$

Range is $[10, 30]$.

ii $g(x) = 10\sin ax + c$

$= 5 \times (2\sin(3 \times 2x) + 4)$

$= 10\sin 6x + 20$

$a = 6, c = 20$

iii $12 = 10\sin 6x + 20$

$-8 = 10\sin 6x$

$\dfrac{-4}{5} = \sin 6x$ where $\pi \le x \le \dfrac{4\pi}{3}$

Hint
Ensure your calculator is in radians mode.

$6x = \pi + 0.9272\ldots, 2\pi - 0.9272\ldots,$
　　　$3\pi + 0.9272\ldots, 4\pi - 0.9272\ldots,$
　　　$5\pi + 0.9272\ldots, 6\pi - 0.9272\ldots,$
　　　$\mathbf{7\pi + 0.9272\ldots, 8\pi - 0.9272\ldots}$

Only the last 2 values will give solutions that will fall in the range $\left[\dfrac{3\pi}{4}, \pi\right]$.

$x = 3.8197\ldots, 4.0342\ldots$

$\approx 3.82, 4.03$

Question 17

$2\cos^2 x = 3\sin x$

$2(1 - \sin^2 x) = 3\sin x$ since $\sin^2 x + \cos^2 x = 1$

$2 - 2\sin^2 x = 3\sin x$

$0 = 2\sin^2 x + 3\sin x - 2$

$= (2\sin x - 1)(\sin x + 2)$

$\sin x = \dfrac{1}{2}, -2$

$\sin x = \dfrac{1}{2}$　　　　$\sin x = -2$

$x = \dfrac{\pi}{6}, \dfrac{5\pi}{6}$　　no solution

So $x = \dfrac{\pi}{6}, \dfrac{5\pi}{6}$.

9780170459228

Question 18

a

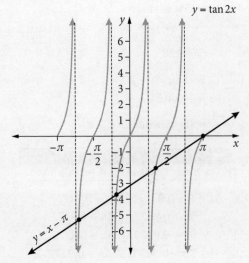

$y = \tan 2x$

b $\tan 2x + \pi = x$

so $\tan 2x = x - \pi$

From the graph in part **a**, there are 4 solutions for $-\pi \le x \le \pi$.

Question 19

a $f(x) = 3\sin\left(2x + \dfrac{\pi}{4}\right) + 2$ for $-\pi \le x \le \pi$ is shown as the curve below.

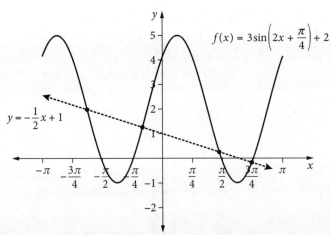

b   The straight line $y = -\dfrac{1}{2}x + 1$ is shown by the dashed line above.

There are 4 solutions for $-\pi \le x \le \pi$.

Question 20

a   Amplitude, $a = \dfrac{35\,000 - 5000}{2}$

               $= 15\,000$

    Vertical intercept, $b = \dfrac{35\,000 + 5000}{2}$

                        $= 20\,000$

b   Comparing the functions, positive gradient = increasing function. This is the graph of $c(t)$:

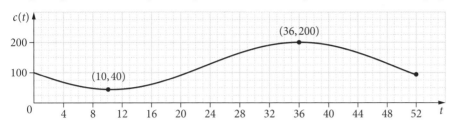

Period $= \dfrac{2\pi}{\dfrac{\pi}{26}} = 52$

Phase: 10 to the right

Reflected vertically from $y = \cos x$, so it has a minimum at $x = 10$ and a maximum at

$x = 10 + \dfrac{1}{2} \times 52 = 36$.

The population of cats increases for $10 < t < 36$.

The graph in the question has 4 sections of width 13, and the population of mice increases for $0 < t < 13$, $39 < t < 52$.

So both cat *and* mouse populations are increasing for $10 < t < 13$.

HSC EXAM TOPIC GRID (2011–2020)

This table shows the coverage of this topic in past HSC exams by question number. The past exams can be downloaded from the NESA website (www.educationstandards.nsw.edu.au) by selecting 'Year 11 – Year 12', 'HSC exam papers'. NESA marking feedback and guidelines can also be found there.

Before 2020, 'Mathematics Advanced' was called 'Mathematics'. For these exams, select 'Year 11 – Year 12', 'Resources archive', 'HSC exam papers archive'.

	Transformations of trigonometric functions	Trigonometric equations	Applications of trigonometric functions
2011		2(b)	
2012		6	
2013	6, 13(a)	13(a)	13(a)
2014		7, 15(a)	
2015		12(a)	15(c)
2016	6	8, 11(g)	
2017	14(a)		
2018		15(a)(iii)	15(a)
2019	7		
2020 new course	**6, 31**		31

Questions in **bold** can be found in this chapter.

CHAPTER 3
DIFFERENTIATION

9780170459228

DIFFERENTIATION

Differentiation rules

$y = x^n$ $\dfrac{dy}{dx} = nx^{n-1}$

$y = [f(x)]^n$ $\dfrac{dy}{dx} = nf'(x)[f(x)]^{n-1}$

Product rule

$y = uv$ $\dfrac{dy}{dx} = u\dfrac{dv}{dx} + v\dfrac{du}{dx}$

Quotient rule

$y = \dfrac{u}{v}$ $\dfrac{dy}{dx} = \dfrac{v\dfrac{du}{dx} - u\dfrac{dv}{dx}}{v^2}$

Chain rule

$y = g(u)$ where $u = f(x)$ $\dfrac{dy}{dx} = \dfrac{dy}{du} \times \dfrac{du}{dx}$

The first derivative

- Stationary point: $\dfrac{dy}{dx} = 0$

- Increasing function: $\dfrac{dy}{dx} > 0$

- Decreasing function: $\dfrac{dy}{dx} < 0$

Stationary points

- A stationary point occurs where $\dfrac{dy}{dx} = 0$

- Maximum turning point if $\dfrac{d^2y}{dx^2} < 0$

- Minimum turning point if $\dfrac{d^2y}{dx^2} > 0$

- Horizontal point of inflection is where
  $\dfrac{dy}{dx} = \dfrac{d^2y}{dx^2} = 0$ and change in concavity occurs.

Derivatives of trigonometric functions

$y = \sin x$ $\dfrac{dy}{dx} = \cos x$

$y = \cos x$ $\dfrac{dy}{dx} = -\sin x$

$y = \tan x$ $\dfrac{dy}{dx} = \sec^2 x$

Derivatives of exponential and logarithmic functions

$y = e^x$ $\dfrac{dy}{dx} = e^x$

$y = e^{f(x)}$ $\dfrac{dy}{dx} = f'(x)e^{f(x)}$

$y = a^x$ $\dfrac{dy}{dx} = (\ln a)a^x$

$y = \ln x$ $\dfrac{dy}{dx} = \dfrac{1}{x}$

$y = \ln f(x)$ $\dfrac{dy}{dx} = \dfrac{f'(x)}{f(x)}$

$y = \log_a x$ $\dfrac{dy}{dx} = \dfrac{1}{x \ln a}$

The second derivative and concavity

- Concave up: $\dfrac{d^2y}{dx^2} > 0$

- Concave down: $\dfrac{d^2y}{dx^2} < 0$

$\dfrac{dy}{dx} = 0$ for point of inflection and check either side for change in concavity.

Optimisation and motion problems

- Maximum and minimum problems
- Displacement, velocity, acceleration

Glossary

chain rule

A rule for differentiating a composite function (function of a function) $y = g(u)$ where $u = f(x)$, or $g[f(x)]$.

$$\frac{dy}{dx} = \frac{dy}{du} \times \frac{du}{dx}$$

concavity

The shape of a curve or part of a curve, determined by the second derivative of its equation. If $f''(x)$ is positive, then the curve is concave up. If $f''(x)$ is negative, then the curve is concave down.

first derivative

The first derivative $f'(x)$ is the rate of change of the function $f(x)$.

global maximum or minimum

The absolute highest or lowest value of a function over a given domain. Differs from **local maximum or minimum**.

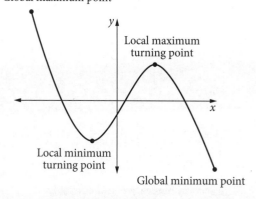

Global maximum point

Local maximum turning point

Local minimum turning point

Global minimum point

horizontal point of inflection

A stationary point on a curve where the graph is flat but the concavity changes.

local maximum or minimum

A relative highest or lowest value of a function over a given domain, indicated graphically by a local turning point or vertex.

normal to a curve

The straight line that is perpendicular to a tangent to the curve at a specific point.

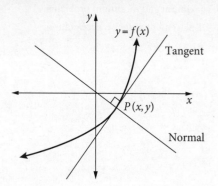

$y = f(x)$

Tangent

$P(x, y)$

Normal

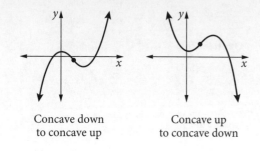

point of inflection

A point on the graph of a function where the concavity changes from concave up to down or concave down to up. $f''(x) = 0$ at a point of inflection and the sign of $f''(x)$ on each side is different.

Concave down to concave up

Concave up to concave down

product rule

A rule for differentiating a product of functions $y = u \times v$.

$$\frac{dy}{dx} = u\frac{dv}{dx} + v\frac{du}{dx}$$

quotient rule

A rule for differentiating a quotient of functions $y = \dfrac{u}{v}$:

$$\frac{dy}{dx} = \frac{v\dfrac{du}{dx} - u\dfrac{dv}{dx}}{v^2}$$

second derivative

The derivative of the derivative of a function, written as $f''(x)$ or $\dfrac{d^2y}{dx^2}$.

sketch

To draw a function that shows important features, such as intercepts and asymptotes, but not draw it precisely to scale.

stationary point

A point on the graph of a function $y = f(x)$ where the gradient is 0, so $f'(x) = 0$. It could be a maximum or minimum turning point or a horizontal point of inflection.

tangent to a curve

A straight line that 'touches' the graph of a function at a specific value of x. Its gradient measures the rate of change of the curve at that point.

Topic summary

Differential calculus (MA-C2)

Tangents and normals (Year 11)

A **tangent to a curve** at a point is a straight line that touches the curve at that point. The gradient of the tangent gives the instantaneous **rate of change** of the function that the curve represents, and is found using the **derivative**.

A **normal to a curve** at a point is a straight line that is perpendicular to the tangent at that point. The gradient of the normal is the **negative reciprocal** of the gradient of the tangent, using the following property:

If 2 lines with gradients m_1 and m_2 are perpendicular, then

$$m_1 m_2 = -1 \text{ or } m_2 = -\frac{1}{m_1}.$$

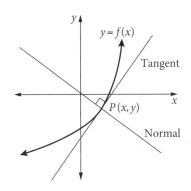

C2.1 Differentiation of trigonometric, exponential and logarithmic functions

The trigonometric functions

$y = \sin x \qquad \dfrac{dy}{dx} = \cos x$

$y = \cos x \qquad \dfrac{dy}{dx} = -\sin x$

$y = \tan x \qquad \dfrac{dy}{dx} = \sec^2 x$

$y = \sin f(x) \qquad \dfrac{dy}{dx} = f'(x)\cos f(x)$

$y = \cos f(x) \qquad \dfrac{dy}{dx} = -f'(x)\sin f(x)$

$y = \tan f(x) \qquad \dfrac{dy}{dx} = f'(x)\sec^2 f(x)$

The exponential functions

$y = e^x \qquad \dfrac{dy}{dx} = e^x$

$y = a^x \qquad \dfrac{dy}{dx} = (\ln a)a^x$

$y = e^{f(x)} \qquad \dfrac{dy}{dx} = f'(x)e^{f(x)}$

$y = a^{f(x)} \qquad \dfrac{dy}{dx} = (\ln a)f'(x)a^{f(x)}$

The logarithmic functions

$y = \ln x \qquad \dfrac{dy}{dx} = \dfrac{1}{x}$

$y = \log_a x \qquad \dfrac{dy}{dx} = \dfrac{1}{x\ln a}$

$y = \ln f(x) \qquad \dfrac{dy}{dx} = \dfrac{f'(x)}{f(x)}$

$y = \log_a f(x) \qquad \dfrac{dy}{dx} = \dfrac{f'(x)}{(\ln a)f(x)}$

> **Hint**
> The rules of differentiation and the '$f(x)$' form of the derivatives appear on the HSC exam reference sheet, which is also printed at the back of this book.

Logarithm laws

(handy for differentiating logarithmic functions)

$\log_a(xy) = \log_a x + \log_a y$

$\log_a\left(\dfrac{x}{y}\right) = \log_a x - \log_a y$

$\log_a(x^n) = n\log_a x$

$\log_a a = 1$

$\log_a 1 = 0$

$\log_a\left(\dfrac{1}{x}\right) = -\log_a x$

Change of base: $\log_a x = \dfrac{\log_b x}{\log_b a}$

$\log_a a^x = x = a^{\log_a x}$

> **Hint**
> The last 2 laws listed appear on the HSC exam reference sheet.

C2.2 Rules of differentiation

Basic differentiation

$y = f(x)^n$ $\qquad \dfrac{dy}{dx} = nf'(x)[f(x)]^{n-1}$

Product rule

$y = uv$ $\qquad \dfrac{dy}{dx} = u\dfrac{dv}{dx} + v\dfrac{du}{dx}$

Quotient rule

$y = \dfrac{u}{v}$ $\qquad \dfrac{dy}{dx} = \dfrac{v\dfrac{du}{dx} - u\dfrac{dv}{dx}}{v^2}$

Chain rule

$y = g(u)$ where $u = f(x)$ $\qquad \dfrac{dy}{dx} = \dfrac{dy}{du} \times \dfrac{du}{dx}$

Applications of differentiation (MA-C3)

C3.1 The first and second derivatives

The derivative of a function is also called the **first derivative**, because the derivative of this derivative is called the **second derivative**.

Notations for the first derivative: $\qquad \dfrac{dy}{dx}, f'(x), y'$

Notations for the second derivative: $\qquad \dfrac{d^2y}{dx^2}, f''(x), y''$

Increasing and decreasing curves

If $f'(x) > 0$, the graph of $y = f(x)$ is increasing (positive gradient).

If $f'(x) < 0$, the graph of $y = f(x)$ is decreasing (negative gradient).

If $f'(x) = 0$, the graph of $y = f(x)$ is stationary (zero gradient).

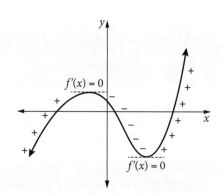

Stationary points

A **stationary point** is a point on the graph of a function $y = f(x)$ where the derivative $f'(x) = 0$.

Here the graph is flat because the gradient is 0. There are three types of stationary points:

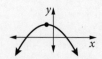

Maximum turning point

Minimum turning point

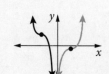

Horizontal point of inflection

A **horizontal point of inflection** is not a turning point. The graph is increasing or decreasing on *both* sides of the inflection point. The graph changes **concavity** at this point.

Stationary points are called **local maximum** or **local minimum** points because they are not necessarily the **global maximum** or **minimum** points on the curve.

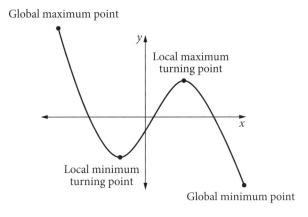

The second derivative and concavity

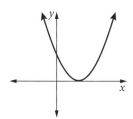

$f''(x) > 0$
concave up

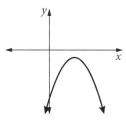

$f''(x) < 0$
concave down

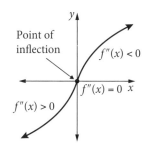

$f''(x) = 0$
concavity is different on each
side of x: point of inflection

Testing for stationary points

A maximum turning point exists when

$$\frac{dy}{dx} = 0 \text{ and } \frac{d^2y}{dx^2} < 0$$

OR

check either side of the stationary point using the first derivative.

x	LHS	Stat. pt	RHS
$\frac{dy}{dx}$	> 0	$= 0$	< 0

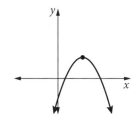

A minimum turning point exists when

$$\frac{dy}{dx} = 0 \text{ and } \frac{d^2y}{dx^2} > 0$$

OR

check either side of the stationary point using the first derivative.

x	LHS	Stat. pt	RHS
$\frac{dy}{dx}$	< 0	$= 0$	> 0

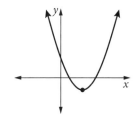

Testing for horizontal points of inflection

$\dfrac{dy}{dx} = 0$ and $\dfrac{d^2y}{dx^2} = 0$ *and* change in concavity occurs; check table below.

x	LHS	POI	RHS
$\dfrac{d^2y}{dx^2}$	< 0	$= 0$	> 0

Concave down to concave up

$\dfrac{dy}{dx} = 0$ and $\dfrac{d^2y}{dx^2} = 0$ *and* change in concavity occurs; check table below.

x	LHS	POI	RHS
$\dfrac{d^2y}{dx^2}$	> 0	$= 0$	< 0

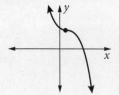

Concave up to concave down

Points of inflection (non-horizontal)

When $\dfrac{dy}{dx} \neq 0$ *and* $\dfrac{d^2y}{dx^2} = 0$ *and* change in concavity occurs; check table below.

x	LHS	POI	RHS
$\dfrac{d^2y}{dx^2}$	< 0	$= 0$	> 0
$\dfrac{d^2y}{dx^2}$	> 0	$= 0$	< 0

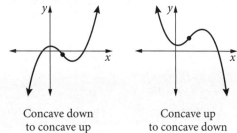

Concave down to concave up

Concave up to concave down

C3.2 Applications of the derivative

Motion in a straight line

Displacement, x

$\downarrow$

Velocity, $v = \dfrac{dx}{dt}$

$\downarrow$

Acceleration, $a = \dfrac{dv}{dt} = \dfrac{d^2x}{dt^2}$

Practice set 1

Multiple-choice questions

Solutions start on page 70.

Question 1

What is the derivative of $y = \log_e(5 + 3x)$?

A $\dfrac{3}{5 + 3x}$ **B** $\dfrac{5}{5 + 3x}$ **C** $5 + \dfrac{1}{x}$ **D** $\dfrac{3}{(5 + 3x)^2}$

Question 2

Find $\dfrac{dy}{dx}$ given that $y = \tan\dfrac{x}{3}$.

A $3\sec^2\dfrac{x}{3}$ **B** $\dfrac{1}{3}\sec^2\dfrac{x}{3}$ **C** $3\sec^2\dfrac{x}{9}$ **D** $\dfrac{1}{9}\sec^2\dfrac{x}{9}$

Question 3

Find the gradient of the tangent to the graph of $y = 2\sin\pi x$ at $x = \dfrac{1}{3}$.

A $\dfrac{\pi}{2}$ **B** π **C** $\sqrt{3}\pi$ **D** 2π

Question 4

The displacement, s metres, at time t seconds of an object moving in a straight line is given by

$$s = t^3 - 6t^2 - 8t - 5.$$

What is the equation of its acceleration, a, in terms of t?

A $a = t^3 - 6t^2 - 8t - 5$ **B** $a = 3t^2 - 12t - 8$ **C** $a = 6t - 12$ **D** $a = 6$

Question 5

Which expression is the derivative of $y = a^{5x}$?

A $5\ln a \times a^{5x}$ **B** $\ln a \times a^{5x}$ **C** $\dfrac{1}{5}\ln a \times a^{5x}$ **D** $\dfrac{a^{5x}}{\ln a}$

Question 6

Given $f(x) = x\sin x$, what is $f''(x)$?

A $\cos x - x\cos x$ **B** $2\cos x - x\sin x$ **C** $x\sin x - 2\cos x$ **D** $x\cos x - 2\sin x$

Question 7

Find the derivative of $y = \dfrac{e^{2x}}{\cos x}$.

A $\dfrac{e^{2x}(2\cos x + \sin x)}{\cos^2 x}$ **B** $\dfrac{e^{2x}(2\cos x - \sin x)}{\cos^2 x}$ **C** $\dfrac{e^{2x}(\cos x - 2\sin x)}{\cos^2 x}$ **D** $\dfrac{e^{2x}(1 + 2\sin x)}{\cos x}$

Question 8

Find the value of the derivative of $y = 3\sin 4x - 4\tan x$ at $x = 0$.

A -16 **B** -4 **C** 1 **D** 8

Question 9

Find the derivative of $\log_e(\cos x)$.

A $\sec x$ **B** $\tan x$ **C** $-\sec x$ **D** $-\tan x$

Question 10 ⬤◯◯

A particle moves in a straight line and its velocity after t seconds is given by $v = 2t^2 - 18$ metres per second.

Find the time at which the particle is at rest.

A $t = 2$ **B** $t = 3$ **C** $t = 8$ **D** $t = 9$

Question 11 ⬤⬤◯

Find the equation of the tangent to the curve $y = 2\log_e x + 1$ at $(e, 3)$.

A $0 = 2x - ey + e$ **B** $0 = 2x - ey + 5e$ **C** $0 = 2x + ey + 3e - 2$ **D** $0 = 2x + ey - (3e + 2)$

Question 12 ⬤⬤◯

Find the derivative of $y = \log_2 3x$.

A $\dfrac{1}{\ln 2}\ln 3x$ **B** $\dfrac{1}{\ln 3}\ln 9x$ **C** $\dfrac{1}{\ln 2} \times \dfrac{1}{x}$ **D** $\dfrac{1}{\ln 3} \times \dfrac{1}{x}$

Question 13 ⬤⬤◯

For what values of x is the graph of the function $y = x^3 - 5x^2 + 4$ concave up?

A $x > 1$ **B** $x < \dfrac{5}{3}$ **C** $x < 1$ **D** $x > \dfrac{5}{3}$

Question 14 ⬤◯◯

Which point on the diagram below satisfies the following?

$$y > 0, \frac{dy}{dx} > 0 \text{ and } \frac{d^2y}{dx^2} < 0$$

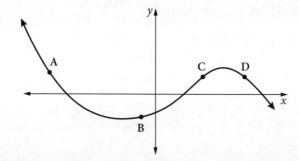

Question 15 ⬤⬤⬤

This diagram shows the graph of $y = f'(x)$.

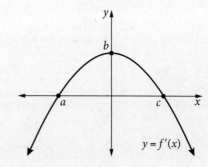

Which statement is true about $y = f(x)$?

A $y = f(x)$ could have a maximum turning point at $x = a$ and a minimum turning point at $x = c$.

B $y = f(x)$ could have a minimum turning point at $x = a$ and a maximum turning point at $x = c$.

C $y = f(x)$ could have a maximum turning point at $x = b$ only.

D $y = f(x)$ could have a minimum turning point at $x = b$ only.

Question 16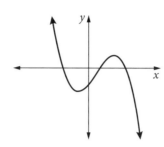

What is the derivative of $\cos(\log_e x)$ when $x > 0$?

A $\dfrac{-1}{x}\sin(\log_e x)$ 　　　　**B** $-\sin\left(\dfrac{\log_e x}{x}\right)$ 　　　　**C** $\sin\dfrac{1}{x}$ 　　　　**D** $\sin(\log_e x)$

Question 17

The graph of the function $y = f(x)$ is shown below.

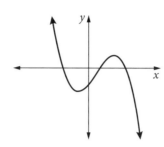

Which one of the following graphs could represent the derivative $y = f'(x)$?

A

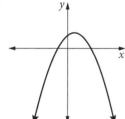

B

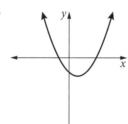

C

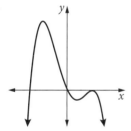

D

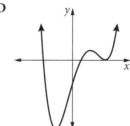

Question 18

Given the function $f(x) = \sin^3 x$, what are the values of a, b and m when $f'(x) = a\cos x + b\cos^m x$?

A $a = b = m = 3$ 　　　**B** $a = b = 3, m = -3$ 　　　**C** $a = m = -3, b = 3$ 　　　**D** $a = m = 3, b = -3$

Question 19

An object moves so that its velocity, v, as a function of time, t, in seconds, is given by $v = 5e^{-t}(1 - 2t)$.

At what time, in seconds, is the acceleration, a, zero?

A $t = 1$ 　　　　**B** $t = 1.5$ 　　　　**C** $t = 2$ 　　　　**D** $t = 2.5$

Question 20

Which statement about the graph of $y = (x - 1)^3(x + 1)$ is *false*?

A It has 2 x-intercepts. 　　　　　　　**B** It has a horizontal inflection point at $x = 1$.

C Its y-intercept is -1. 　　　　　　　**D** It has a maximum turning point at $x = -\dfrac{1}{2}$.

Practice set 2

Short-answer questions

Solutions start on page 72.

Question 1 (5 marks) ●○○

Differentiate each function with respect to x.

a $y = e^{2x}$ 1 mark

b $y = 3e^{\sin x}$ 1 mark

c $y = \ln(e^x + 1)$ 1 mark

d $y = 6\cos\dfrac{x}{2}$ 2 marks

Question 2 (4 marks) ●○○

Consider the equation $f(x) = \cos 2x$.

a Find:

   i $f'(x)$ 1 mark

   ii $f''(x)$ 1 mark

b Hence, evaluate:

   i $f'\left(\dfrac{\pi}{3}\right)$ 1 mark

   ii $f''\left(\dfrac{\pi}{2}\right)$ 1 mark

Question 3 (2 marks) ●●○

For what values of x is the curve $y = 12x - 3x^2$ decreasing?   2 marks

Question 4 (6 marks) ●●●

Differentiate each function with respect to x.

a $f(x) = (1 + \tan x)^2$ 2 marks

b $f(x) = x\log_e x$ 2 marks

c $f(x) = \sqrt{\sin x}$ 2 marks

Question 5 (4 marks) ●○○

Consider the equation $y = x^3 - 3x^2 + 8x - 2$.

a Find $\dfrac{dy}{dx}$ and $\dfrac{d^2y}{dx^2}$.   2 marks

b Hence, determine the value of x for which $\dfrac{d^2y}{dx^2} = 0$.   2 marks

Question 6 (3 marks) ⚫⚫⚪

Consider $y = \log_e\left(\dfrac{2+x}{2-x}\right)$.

a Expand $\log_e\left(\dfrac{2+x}{2-x}\right)$ using a logarithm law.   1 mark

b Hence, show that $\dfrac{d}{dx}\log_e\left(\dfrac{2+x}{2-x}\right) = \dfrac{4}{4-x^2}$.   2 marks

Question 7 (2 marks) ⚫⚫⚪

Given $y = \cos^2 x$, find the exact value of the gradient of the tangent at $x = \dfrac{\pi}{4}$.   2 marks

Question 8 (6 marks) ⚫⚫⚪

Differentiate:

a $y = \ln(x\sqrt{x})$ 2 marks

b $y = x^2 \times 4^x$ 2 marks

c $y = \dfrac{\log_2 x}{x^2}$ 2 marks

Question 9 (8 marks) ⚫⚫⚪

Consider the graph of the function $y = 3x^2 - x^3$.

a Find the stationary points and determine their nature.   2 marks

b Given that the point $(1, 2)$ lies on the curve, show that it is a point of inflection.   2 marks

c Sketch the curve, clearly showing the stationary points, the inflection point and the coordinates of any x- and y-intercepts.   3 marks

d For what values of x is the curve concave up?   1 mark

Question 10 (2 marks) ⚫⚫⚪

Calculate the gradient of the normal to the function $y = e^{-x}$ at $x = \ln 2$.   2 marks

Question 11 (8 marks) ⚫⚫⚪

Consider the graph of $y = x^3 - x^2 - 8x + 5$.

a Find the stationary points and determine their nature.   4 marks

b Given that point $A\left(\dfrac{1}{3}, \dfrac{61}{27}\right)$ lies on the curve, show that there is an inflection point at A.   2 marks

c Sketch the curve, clearly indicating the y-intercept, the stationary points and the point of inflection.   2 marks

Question 12 (2 marks) ⚫⚫⚫

Show that $y = \ln\dfrac{1}{x}$ is decreasing for all values of x.   2 marks

PRACTICE SET 2

Question 13 (5 marks) ●●▮

The position, in centimetres, of a particle moving in a straight line is given by $x = t - 1 - 5\ln(t + 1)$, where t is in seconds.

a   Find the initial displacement of the particle.   1 mark

b   Find the time when the particle is at rest.   2 marks

c   What is the acceleration at $t = 4$?   2 marks

Question 14 (3 marks) ●●●

Find the coordinates of the stationary point on the curve $y = \dfrac{4x}{\sqrt{x - 1}}$ and determine its nature.   3 marks

Question 15 (2 marks) ●●▮

Show that $\dfrac{d}{dx}(\operatorname{cosec} x) = -\cot x \operatorname{cosec} x$.   2 marks

Question 16 (3 marks) ●●●

The function $y = ax^4 - 4x^3 + bx^2$ has a stationary point at $(2, -32)$.

Determine the values of a and b.   3 marks

Question 17 (3 marks) ●●●

Find the coordinates of the point on the graph of $y = \tan x$, $0 \le x < \dfrac{\pi}{2}$, where the normal at this point has a gradient of $-\dfrac{1}{2}$.   3 marks

Question 18 (6 marks) ●●●

The number of people, $P(t)$, infected by a virus in a city, t days after the first cases of infection were detected, can be modelled by

$$P(t) = 5 + 18te^{-0.12t}.$$

a   How many people in this city were:

   i   initially infected with the virus?   1 mark

   ii   diagnosed with the virus on the fifth day?   1 mark

b   On which day were the maximum number of infections recorded in this city?   2 marks

c   What is the rate of change of the number of people being infected by the virus in this city on the 20th day? Answer correct to one decimal place.   2 marks

Question 19 (7 marks) ●●■

A farmer wants to make three square enclosures, as shown in the diagram below. He uses an existing L-shaped wall for some of the sides. He uses 36 metres of fencing for all the other sides, as shown, with two of the squares being the same size. Sides x and y are measured in metres.

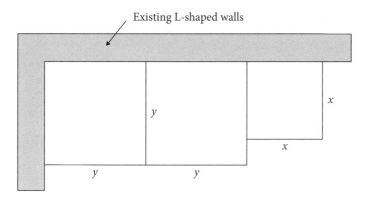

Existing L-shaped walls

a **i** Show that $y = \dfrac{18 - x}{2}$.   2 marks

    ii Hence, show that the total area of the enclosure is represented by $A = \dfrac{(18 - x)^2 + 2x^2}{2}$.   2 marks

b Hence, calculate the value of x for which the total area of the enclosures is a minimum and find the minimum total area for the three enclosures.   3 marks

Question 20 (7 marks) ●●●

A drone, D, has crashed in difficult terrain and Phoebe is driving to its location from point P. From the crashed drone to the nearest point on a straight road, Q, is 4 kilometres and from there it is 10 kilometres to point P. Phoebe drives at 30 km/h over the difficult terrain and 50 km/h along the straight road. Phoebe decides to turn off the straight road at X, x kilometres from Q.

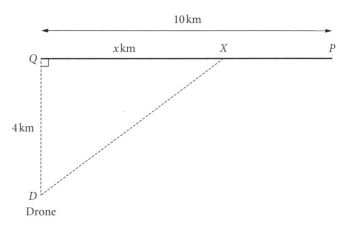

a If t represents the time in hours Phoebe takes to travel from P to D, show that   3 marks

$$t = \frac{\sqrt{x^2 + 16}}{30} + \frac{10 - x}{50}.$$

b Find the value of x that will give the quickest route for Phoebe to take. Write your answer in exact form.   2 marks

c Determine the time taken, in minutes, for Phoebe to drive the quickest route. Answer correct to one decimal place.   2 marks

Practice set 1

Worked solutions

1 A

$$y = \log_e(5 + 3x)$$

$$\frac{dy}{dx} = \frac{3}{5 + 3x}$$

2 B

$$y = \tan\frac{x}{3}$$

$$\frac{dy}{dx} = \frac{1}{3}\sec^2\frac{x}{3}$$

3 B

$$y = 2\sin\pi x$$

$$\frac{dy}{dx} = 2\pi\cos\pi x$$

When $x = \frac{1}{3}$,

$$\frac{dy}{dx} = 2\pi\cos\frac{\pi}{3}$$

$$= \cancel{2}\pi \times \frac{1}{\cancel{2}}$$

$$= \pi$$

4 C

$$s = t^3 - 6t^2 - 8t - 5$$

$$v = \frac{ds}{dt} = 3t^2 - 12t - 8$$

$$a = \frac{dv}{dt} = 6t - 12$$

5 A

6 B

$$f(x) = x\sin x$$

$$f'(x) = x\cos x + \sin x$$

$$f''(x) = x(-\sin x) + \cos x + \cos x$$

$$= 2\cos x - x\sin x$$

7 A

$$y = \frac{e^{2x}}{\cos x}$$

$$y' = \frac{\cos x \times 2e^{2x} - e^{2x}(-\sin x)}{\cos^2 x}$$

$$= \frac{2e^{2x}\cos x + e^{2x}\sin x}{\cos^2 x}$$

$$= \frac{e^{2x}(2\cos x + \sin x)}{\cos^2 x}$$

8 D

$$y = 3\sin 4x - 4\tan x$$

$$y' = 12\cos 4x - 4\sec^2 x$$

When $x = 0$,

$$y' = 12 \times 1 - 4 \times 1 = 8$$

9 D

$$y = \log_e(\cos x)$$

$$y' = \frac{-\sin x}{\cos x}$$

$$= -\tan x$$

10 B

The particle is at rest when $v = 0$.

$$0 = 2t^2 - 18$$

$$= 2(t^2 - 9)$$

$$= 2(t + 3)(t - 3)$$

$$t = 3, -3$$

But $t > 0$, so $t = 3$ only.

11 A

$$y = 2\ln x + 1$$

$$y' = \frac{2}{x}$$

When $x = e$, $y' = \frac{2}{e}$.

$$y - 3 = \frac{2}{e}(x - e)$$

$$ey - 3e = 2x - 2e$$

$$0 = 2x - ey + e$$

12 C

$$y = \frac{\log_e 3x}{\log_e 2} = \frac{1}{\ln 2}\log_e 3x$$

$$y' = \frac{1}{\ln 2} \times \frac{\cancel{3}}{\cancel{3}x}$$

$$= \frac{1}{\ln 2} \times \frac{1}{x}$$

13 D

$$y = x^3 - 5x^2 + 4$$

$$y' = 3x^2 - 10x$$

$$y'' = 6x - 10$$

$$6x - 10 > 0$$

$$6x > 10$$

$$x > \frac{10}{6}$$

$$> \frac{5}{3}$$

14 C

y > 0, as the point is above the x-axis, $\dfrac{dy}{dx} > 0$,

as the curve is increasing, and $\dfrac{d^2y}{dx^2} < 0$ as it is

concave down.

15 B

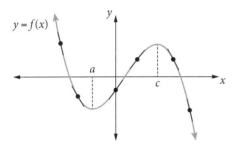

16 A

$y = \cos(\ln x)$

$y' = -\dfrac{1}{x}\sin(\ln x)$

17 A

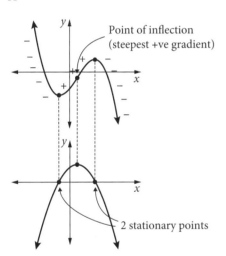

18 D

$f(x) = (\sin x)^3$

$f'(x) = 3\sin^2 x \cos x$

$\quad = 3(1 - \cos^2 x)\cos x$

$\quad = 3\cos x - 3\cos^3 x$

$a = 3, b = -3, m = 3$

19 B

$v = 5e^{-t}(1 - 2t)$

$a = \dfrac{dv}{dt} = 5e^{-t}(-2) + (1 - 2t)(-5e^{-t})$

$\quad = -5e^{-t}(2 + 1 - 2t)$

$0 = -5e^{-t}(3 - 2t)$

$3 - 2t = 0$

$\quad t = \dfrac{3}{2} = 1.5\,\text{s}$

20 D

$y = (x - 1)^3(x + 1)$

x-intercept: $0 = (x - 1)^3(x + 1)$

$x = 1, -1$ A is true.

y-intercept: $y = -1 \times 1 = -1$

$(0, -1)$ C is true.

$y' = (x - 1)^3 + 3(x + 1)(x - 1)^2$

$\quad = (x - 1)^2(x - 1 + 3x + 3)$

$\quad = (x - 1)^2(4x + 2)$

$\quad = 2(x - 1)^2(2x + 1)$

$y' = 0$ (stationary points)

$0 = 2(x - 1)^2(2x + 1)$

$x = 1, -\dfrac{1}{2}$

$y'' = 2(x - 1)^2 \times 2 + 2(2x + 1) \times 2(x - 1)$

$\quad = 4(x - 1)^2 + 4(2x + 1)(x - 1)$

$\quad = 4(x - 1)[x - 1 + 2x + 1]$

$\quad = 4(x - 1)(3x)$

$\quad = 12x(x - 1)$

When $x = 1, y'' = 12 \times (1 - 1) = 0$

so a possible horizontal inflection point.

When $x = -\dfrac{1}{2}$,

$y'' = 12 \times -\dfrac{1}{2}\left(-\dfrac{1}{2} - 1\right)$

$\quad = -6 \times -\dfrac{3}{2}$

$\quad = 9 > 0$ minimum turning point

D is not correct.

$x = 1$ is a possible horizontal inflection point.

Check change in concavity in table below:

x	0.9	1	1.1
y''	$-1.08 < 0$	$= 0$	$1.32 > 0$

Since change in concavity occurs and $y' = y'' = 0$, $x = 1$ is a horizontal inflection point.

B is correct.

Practice set 2

Worked solutions

Question 1

a $y' = 2e^{2x}$

b $y' = 3\cos x e^{\sin x}$

c $y' = \dfrac{e^x}{e^x + 1}$

d $y' = -6 \times \dfrac{1}{2}\sin\dfrac{x}{2}$

$\qquad = -3\sin\dfrac{x}{2}$

Question 2

a **i** $f'(x) = -2\sin 2x$

   ii $f''(x) = -4\cos 2x$

b **i** $f'\left(\dfrac{\pi}{3}\right) = -2\sin\dfrac{2\pi}{3}$

$\qquad\qquad = -2 \times \dfrac{\sqrt{3}}{2}$

$\qquad\qquad = -\sqrt{3}$

   ii $f''\left(\dfrac{\pi}{2}\right) = -4\cos\dfrac{2\pi}{2}$

$\qquad\qquad = -4 \times \cos\pi$

$\qquad\qquad = -4 \times -1$

$\qquad\qquad = 4$

Question 3

$y = 12x - 3x^2$

$y' < 0$ for decreasing curve

$y' = 12 - 6x < 0$

$\qquad 12 < 6x$

$\qquad\quad x > 2$

Question 4

a $f'(x) = 2(1 + \tan x)\sec^2 x$

b $f'(x) = x \times \dfrac{1}{x} + \ln x$

$\qquad = 1 + \ln x$

c $f(x) = (\sin x)^{\frac{1}{2}}$

$\quad f'(x) = \dfrac{1}{2}(\sin x)^{-\frac{1}{2}}\cos x$

$\qquad\quad = \dfrac{\cos x}{2\sqrt{\sin x}}$

Question 5

a $y = x^3 - 3x^2 + 8x - 2$

$\dfrac{dy}{dx} = 3x^2 - 6x + 8$

$\dfrac{d^2y}{dx^2} = 6x - 6$

b $\dfrac{d^2y}{dx^2} = 0$

$6x - 6 = 0$

$\qquad 6x = 6$

$\qquad\; x = 1$

Question 6

a $\log_e(2 + x) - \log_e(2 - x)$

b $\dfrac{d}{dx}\big(\log_e(2 + x) - \log_e(2 - x)\big) = \dfrac{1}{2 + x} - \dfrac{(-1)}{2 - x}$

$\qquad\qquad\qquad\qquad = \dfrac{1}{2 + x} + \dfrac{1}{2 - x}$

$\qquad\qquad\qquad\qquad = \dfrac{2 - x + 2 + x}{(2 + x)(2 - x)}$

$\qquad\qquad\qquad\qquad = \dfrac{4}{4 - x^2}$

Question 7

$y = (\cos x)^2$

$y' = 2(\cos x)(-\sin x)$

$\quad = -2\cos x\sin x$

When $x = \dfrac{\pi}{4}$,

$y' = -2\cos\dfrac{\pi}{4}\sin\dfrac{\pi}{4}$

$\quad = -2 \times \dfrac{1}{\sqrt{2}} \times \dfrac{1}{\sqrt{2}}$

$\quad = -2 \times \dfrac{1}{2} = -1$

The gradient of the tangent is -1.

Question 8

a $y = \ln(x\sqrt{x})$

$\quad = \ln\left(x^{\frac{3}{2}}\right)$

$\dfrac{dy}{dx} = \dfrac{\frac{3}{2}x^{\frac{1}{2}}}{x^{\frac{3}{2}}}$

$\quad\quad = \dfrac{3}{2x}$

b $y = x^2 \times 4^x$

$y' = x^2 \times \ln 4 \times 4^x + 4^x \times 2x$

$\quad = x4^x(x\ln 4 + 2)$

c $y = \dfrac{\log_2 x}{x^2} = \dfrac{\left(\dfrac{\log_e x}{\log_e 2}\right)}{x^2} = \dfrac{\left(\dfrac{1}{\ln 2} \times \ln x\right)}{x^2}$

$y' = \dfrac{x^2 \times \dfrac{1}{\ln 2} \times \dfrac{1}{x} - \left(\dfrac{1}{\ln 2} \times \ln x\right) \times 2x}{x^4}$

$\quad = \dfrac{\dfrac{x}{\ln 2} - \dfrac{2x\ln x}{\ln 2}}{x^4}$

$\quad = \dfrac{\dfrac{\cancel{x}}{\ln 2}(1 - 2\ln x)}{x^{\cancel{4}3}}$

$\quad = \dfrac{1 - 2\ln x}{x^3 \ln 2}$

Question 9

a $y' = 6x - 3x^2$

$y' = 0$ (stationary points)

$0 = 3x(2 - x)$

$x = 0, 2$

When $x = 0$, $y = 0$.

When $x = 2$, $y = 3 \times 2^2 - 2^3 = 12 - 8 = 4$.

$y'' = 6 - 6x$

At $(0,0)$ $y'' = 6 - 0 = 6 > 0$.

This is a minimum turning point.

At $(2,4)$ $y'' = 6 - 6 \times 2 = 6 - 12 = -6 < 0$.

This is a maximum turning point.

b $y'' = 0$ (inflection points)

$6 - 6x = 0$

$6x = 6$

$x = 1$

$y = 3 - 1 = 2$

x	0.9	1	1.1
y''	$0.6 > 0$	$= 0$	$-0.6 < 0$

Since a change in concavity occurs, $(1,2)$ is an inflection point.

c To find x-intercepts:

$3x^2 - x^3 = 0$

$x^2(3 - x) = 0$

$x = 0$ and $x = 3$

So $(0,0)$ and $(3,0)$ are the x-intercepts.

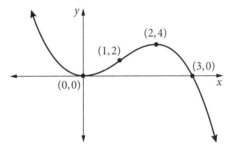

d Curve is concave up when $y'' > 0$.
From graph, that is for $x < 1$.

Question 10

$y = e^{-x}$

$y' = -e^{-x}$

When $x = \ln 2$,

$y' = -e^{-\ln 2}$

$\quad = -e^{\ln 2^{-1}}$

$\quad = 2^{-1}$

$\quad = -\dfrac{1}{2}$

$m_T \times m_N = -1$ (perpendicular lines)

$-\dfrac{1}{2} \times m_N = -1$

$m_N = 2$

The gradient of the normal is 2.

Question 11

a $y = x^3 - x^2 - 8x + 5$

$y' = 3x^2 - 2x - 8$

$y' = 0$ (stationary points)

$0 = (3x + 4)(x - 2)$

$x = -\dfrac{4}{3}, 2$

$y'' = 6x - 2$

When $x = -\dfrac{4}{3}, y = \dfrac{311}{27}$.

$y'' = 6 \times \dfrac{-4}{3} - 2$

$= -8 - 2$

$= -10 < 0$

$\left(-\dfrac{4}{3}, \dfrac{311}{27}\right)$ is a maximum turning point.

When $x = 2, y = -7$.

$y'' = 6 \times 2 - 2$

$= 12 - 2$

$= 10 > 0$

$(2, -7)$ is a minimum turning point.

b $y'' = 0$ (points of inflection)

$6x - 2 = 0$

$6x = 2$

$x = \dfrac{1}{3}$

When $x = \dfrac{1}{3}$,

$y = \left(\dfrac{1}{3}\right)^3 - \left(\dfrac{1}{3}\right)^2 - 8 \times \dfrac{1}{3} + 5$

$= \dfrac{61}{27}$

x	0.3	$\frac{1}{3}$	0.35
y''	$-0.2 < 0$	$= 0$	$0.1 > 0$

There is a change in concavity at $x = \dfrac{1}{3}$, so A is an inflection point.

c

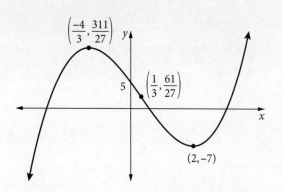

Question 12

$y = \ln\dfrac{1}{x} = \ln x^{-1}$

$y' = \dfrac{-x^{-2}}{x^{-1}} = \dfrac{-1}{x^2} \times \dfrac{x}{1} = \dfrac{-1}{x}$

The logarithm exists for $x > 0$.

$\dfrac{1}{x} > 0$

$\dfrac{-1}{x} < 0$

$\dfrac{dy}{dx} < 0$

So $y = \ln\dfrac{1}{x}$ is decreasing for all x.

Question 13

$x = t - 1 - 5\ln(t + 1)$

a $t = 0, x = 0 - 1 - 5\ln 1$

$= -1 - 5 \times 0$

$= -1$

The particle is initially 1 cm left of the origin.

b $v = \dfrac{dx}{dt} = 1 - 5 \times \dfrac{1}{t + 1}$

$= 1 - \dfrac{5}{t + 1}$

When $v = 0$,

$0 = 1 - \dfrac{5}{t + 1}$

$\dfrac{5}{t + 1} = 1$

$5 = t + 1$

$4 = t$

$t = 4\,\text{s}$

The particle is at rest at 4 seconds.

c $a = \dfrac{dv}{dt}$

$v = 1 - 5(t + 1)^{-1}$

$a = 5(t + 1)^{-2}$

$= \dfrac{5}{(t + 1)^2}$

When $t = 4$,

$a = \dfrac{5}{(4 + 1)^2} = \dfrac{5}{25} = \dfrac{1}{5}$

$= 0.2\,\text{cm/s}^2$

9780170459228

WORKED SOLUTIONS

Question 14

$$y = \frac{4x}{\sqrt{x-1}} = \frac{4x}{(x-1)^{\frac{1}{2}}}$$

$$y' = \frac{(x-1)^{\frac{1}{2}} \times 4 - 4x \times \frac{1}{2}(x-1)^{-\frac{1}{2}}}{\left((x-1)^{\frac{1}{2}}\right)^2}$$

$$= \frac{4\sqrt{x-1} - \frac{2x}{\sqrt{x-1}}}{x-1}$$

$$= \frac{4(x-1) - 2x}{(x-1)^{\frac{3}{2}}}$$

$$= \frac{2x-4}{(x-1)^{\frac{3}{2}}}$$

$$= \frac{2(x-2)}{(x-1)^{\frac{3}{2}}}$$

$y' = 0$ (stationary points)

$$0 = \frac{2(x-2)}{(x-1)^{\frac{3}{2}}}$$

$$= 2(x-2)$$

$x = 2$

When $x = 2$,

$$y = \frac{4 \times 2}{\sqrt{2-1}}$$

$$= 8$$

At $(2, 8)$:

x	1.9	2	2.1
y'	< 0	$= 0$	> 0

$x = 2$

Minimum turning point at $(2, 8)$.

Question 15

$$y = \operatorname{cosec} x = \frac{1}{\sin x} = (\sin x)^{-1}$$

$$y' = -(\sin x)^{-2}\cos x$$

$$= \frac{-\cos x}{\sin^2 x}$$

$$= \frac{-\cos x}{\sin x} \times \frac{1}{\sin x}$$

$$= -\cot x \operatorname{cosec} x$$

Question 16

$$y = ax^4 - 4x^3 + bx^2$$

Stationary point at $(2, -32)$.

$$-32 = 16a - 32 + 4b$$
$$0 = 16a + 4b$$
$$= 4a + b \qquad [1]$$

$$\frac{dy}{dx} = 4ax^3 - 12x^2 + 2bx$$

$$\frac{dy}{dx} = 0 \text{ at } x = 2$$

$$0 = 4a \times 2^3 - 12 \times 2^2 + 2b \times 2$$
$$= 32a - 48 + 4b$$
$$= 8a - 12 + b$$
$$8a + b = 12 \qquad [2]$$

$[2] - [1]$

$$4a = 12$$
$$a = 3$$
$$b = -12$$

Question 17

$$y = \tan x$$

$$\frac{dy}{dx} = \sec^2 x$$

$$m_N = -\frac{1}{2}$$

$$m_T = 2$$

$$2 = \sec^2 x$$

$$\pm\sqrt{2} = \sec x \quad \text{but} \quad 0 \leq x < \frac{\pi}{2}$$

$$\sqrt{2} = \frac{1}{\cos x}$$

$$\cos x = \frac{1}{\sqrt{2}}$$

$$x = \frac{\pi}{4}$$

When $x = \frac{\pi}{4}$, $y = \tan\frac{\pi}{4} = 1$.

$$\left(\frac{\pi}{4}, 1\right)$$

Question 18

a   i $t = 0$, $P(0) = 5 + 18 \times 0 = 5$ people

   ii $t = 5$, $P(5) = 5 + 18 \times 5e^{-12 \times 5} = 54.393\ldots$
$$= 54 \text{ people}$$

b $P'(t) = 18t \times -0.12te^{-0.12t} + e^{-0.12t} \times 18$

$= -2.16te^{-0.12t} + 18e^{-0.12t}$

$= e^{-0.12t}(18 - 2.16t)$

$P'(t) = 0$

$0 = e^{-0.12t}(18 - 2.16t)$

$e^{-0.12t} \neq 0$

$18 - 2.16t = 0$

$2.16t = 18$

$t = \dfrac{18}{2.16}$

$= 8\tfrac{1}{3}$

Maximum number of infections on 8th day.

c $t = 20$

$P'(20) = e^{-0.12 \times 20} \times -25.2$

$= -2.286\ldots$

This is a decrease of 2.3 people per day.

Question 19

a i Length of fencing $= 4y + 2x = 36$

$2y + x = 18$

$2y = 18 - x$

$y = \dfrac{18 - x}{2}$

ii $A = 2y^2 + x^2$

$= 2\left(\dfrac{18 - x}{2}\right)^2 + x^2$

$= 2\dfrac{(18 - x)^2}{4} + x^2$

$= \dfrac{(18 - x)^2 + 2x^2}{2}$

b $A = \dfrac{324 - 36x + x^2 + 2x^2}{2}$

$= \dfrac{324 - 36x + 3x^2}{2}$

$\dfrac{dA}{dx} = -18 + 2 \times \dfrac{3}{2}x$

$= 3x - 18$

$3x - 18 = 0$

$3x = 18$

$x = 6\,\text{m}$

$\dfrac{d^2A}{dx^2} = 3 > 0$ so minimum area

$A = \dfrac{(18 - 6)^2 + 2 \times 6^2}{2} = 108\,\text{m}^2$

Question 20

a

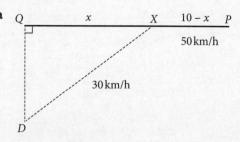

Time $= \dfrac{\text{Distance}}{\text{Speed}}$

Straight road:

$PX = 10 - x$

$t_1 = \dfrac{10 - x}{50}$

In rough terrain:

$XD = \sqrt{4^2 + x^2} = \sqrt{16 + x^2}$

$t_2 = \dfrac{\sqrt{16 + x^2}}{30}$

Total $t = \dfrac{10 - x}{50} + \dfrac{\sqrt{16 + x^2}}{30}$

b $\dfrac{dt}{dx} = -\dfrac{1}{50} + \dfrac{1}{30} \times \dfrac{1}{2}(16 + x^2)^{-\frac{1}{2}} \times 2x$

$0 = -\dfrac{1}{50} + \dfrac{x}{30\sqrt{16 + x^2}}$

$\dfrac{x}{30\sqrt{16 + x^2}} = \dfrac{1}{50}$

$50x = 30\sqrt{16 + x^2}$

$(50x)^2 = 900(16 + x^2)$

$2500x^2 = 14\,400 + 900x^2$

$1600x^2 = 14\,400$

$x^2 = \dfrac{14\,400}{1600}$

$= 9$

$x = \sqrt{9} = 3$ where $x > 0$

$\dfrac{d^2t}{dx^2} = \dfrac{x}{30} \times \dfrac{-1}{2}(16 + x^2)^{\frac{-3}{2}} \times 2x + (16 + x^2)^{\frac{-1}{2}} \times \dfrac{1}{30}$

$= \dfrac{-x^2}{30\sqrt{(16 + x^2)^3}} + \dfrac{1}{30\sqrt{16 + x^2}}$

$= \dfrac{-x^2 + 16 + x^2}{30\sqrt{(16 + x^2)^3}}$

$= \dfrac{16}{30\sqrt{(16 + x^2)^3}} > 0$

Minimum, so it is the quickest route when $x = 3\,\text{km}$.

c $t = \dfrac{\sqrt{3^2 + 4^2}}{30} + \dfrac{10 - 3}{50}$

$\quad = \dfrac{5}{30} + \dfrac{7}{50}$

$\quad = \dfrac{46}{150}$

$\quad \approx 0.307\,\text{h}$

$\quad \approx 18.4 \text{ minutes}$

HSC exam topic grid (2011–2020)

This table shows the coverage of this topic in past HSC exams by question number. The past exams can be downloaded from the NESA website (www.educationstandards.nsw.edu.au) by selecting 'Year 11 – Year 12', 'HSC exam papers'. NESA marking feedback and guidelines can also be found there.

Before 2020, 'Mathematics Advanced' was called 'Mathematics'. For these exams, select 'Year 11 – Year 12', 'Resources archive', 'HSC exam papers archive'.

	Differentiation rules, tangents and normals	Trigonometric, exponential and logarithmic functions	Stationary points, concavity and curve sketching	Optimisation* and motion problems
2011	2(c)–(d), 4(a), 4(d)(i)	1(d), 2(d), 4(a)	7(a), 9(c)	7(b), 9(b)(i), 10(b)*
2012	11(c)–(d), 12(a)	11(d), 12(a), 14(c)	4, 14(a)	15(b)(ii), 16(b)*
2013	4, 11(c)–(d)	4, 11(c)–(d), 16(b)	8, 12(a)	10, 14(b)*
2014	11(c)	13(a)(i), 13(b), 14(a), 15(c)	9, 14(a), 14(e)	9, 13(c), 16(c)*
2015	11(e)–(f), 12(c), 12(e)(i)	6, 11(e)–(f), 15(a), 15(c)(i), 15(c)(iii)	13(c)	15(c)(iii)*, 16(c)*
2016	5, 11(b), 11(f), 12(d)(i), 16(b)	5, 11(f), 12(d)(i), 16(b)	13(a)	14(c)*, 16(a)(i)–(iii), 16(b)*
2017	3, 11(c)–(d), 12(a)	3, 11(c)–(d), 14(c)(i)	4, 9, 13(b)	10, 16(a)*
2018	5, 11(f)–(g), 12(b)	5, 11(f)–(g), 12(b)	9, 13, 14(c)	12(d), 16(a)*
2019	11(b)–(c), 13(c)(i), 16(c)(i)	11(b), 12(c)(ii), 13(c)(i)	8, 14(b)(i), 14(b)(iv)	8, 10, 15(c)*
2020 new course	10, 18(a), 21(b), 29	18(a), 21(b), 29, 31(c)	8, 10, 16	25*

CHAPTER 4
INTEGRATION

9780170459228

INTEGRATION

Anti-differentiation

- Opposite of differentiation
- Anti-derivative, primitive
- The indefinite integral $\int f(x)\,dx$:

$$\int x^n\,dx = \frac{1}{n+1}x^{n+1} + c$$

- The reverse chain rule:

$$\int f'(x)\big[f(x)\big]^n\,dx = \frac{1}{n+1}\big[f(x)\big]^{n+1} + c$$

Exponential functions

$$\int e^x\,dx = e^x + c$$

$$\int e^{ax+b}\,dx = \frac{1}{a}e^{ax+b} + c$$

$$\int a^x\,dx = \frac{1}{\ln a}a^x + c$$

The trapezoidal rule

$$\int_a^b f(x)\,dx \approx \frac{b-a}{2n}\Big\{f(a) + f(b) + 2\big[f(x_1) + \cdots + f(x_{n-1})\big]\Big\}$$

Areas between curves

$$A = \int_a^b \big[f(x) - g(x)\big]\,dx$$

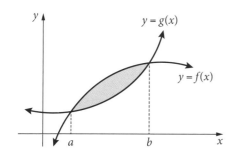

Trigonometric functions

$$\int \sin(ax + b)\,dx = -\frac{1}{a}\cos(ax + b) + c$$

$$\int \cos(ax + b)\,dx = \frac{1}{a}\sin(ax + b) + c$$

$$\int \sec^2(ax + b)\,dx = \frac{1}{a}\tan(ax + b) + c$$

Logarithmic functions

$$\int \frac{1}{x}\,dx = \ln|x| + c$$

$$\int \frac{f'(x)}{f(x)}\,dx = \ln|f(x)| + c$$

Area under a curve

The definite integral $\int_a^b f(x)\,dx$

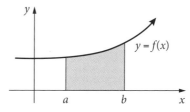

Applications of integration

- Given $f'(x)$ and an initial condition $f(a) = b$, find $f(x)$.
- Problems involving displacement, velocity, acceleration and rates of change.

Glossary

anti-derivative (or primitive or integral)

The opposite of a derivative function. For a function $f(x)$, the anti-derivative is the function $F(x)$ whose derivative is $f(x)$, that is, $F'(x) = f(x)$.

anti-differentiation (or integration)

The process of finding the anti-derivative.

definite integral

An integral that has a value, describing the area under the graph of $y = f(x)$ between $x = a$ and $x = b$, with notation $\int_a^b f(x)\,dx$.

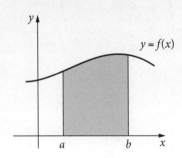

indefinite integral

The general anti-derivative

$$\int f(x)\,dx = F(x) + c,$$

where c is a constant.

limits of integration

The boundary values of a and b for the interval $[a, b]$ in the definite integral $\int_a^b f(x)\,dx$.

reverse chain rule

The integration rule

$$\int f'(x)\big[f(x)\big]^n dx = \frac{1}{n+1}\big[f(x)\big]^{n+1} + c,$$

which is the opposite of the chain rule for differentiation:

$$\frac{d}{dx}[f(x)]^n = n[f(x)]^{n-1} f'(x).$$

trapezoidal rule

The rule

$$\int_a^b f(x)\,dx \approx \frac{h}{2}\big\{f(a) + f(b) + 2\big[f(x_1) + \cdots + f(x_{n-1})\big]\big\},$$

where $h = \dfrac{b-a}{n}$ for approximating the area under the graph of $y = f(x)$ using trapeziums.

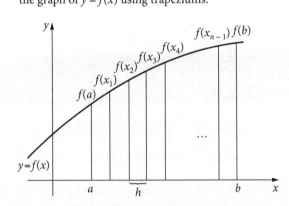

Topic summary

Integral calculus (MA-C4)

C4.1 The anti-derivative

Anti-differentiation

The **anti-derivative** (or **primitive** or **integral**) of $y = f(x)$ is the function $F(x)$ whose derivative is $f(x)$, that is, $F'(x) = f(x)$. If $f(x)$ is the derivative of $F(x)$, then $F(x)$ is the anti-derivative of $f(x)$.

The **indefinite integral** has the notation $F(x) = \int f(x)\,dx$,

$$\int x^n dx = \frac{1}{n+1}x^{n+1} + c, \text{ for } n \neq -1.$$

Example 1

a $\int 6x^2 - x + 3\,dx = 2x^3 - \dfrac{x^2}{2} + 3x + c$ **b** $\int \dfrac{1}{x^2}dx = -\dfrac{1}{x} + c$

> **Hint**
> Don't forget '+ c'.

The reverse chain rule

$$\int (ax + b)^n dx = \frac{1}{a(n+1)}(ax+b)^{n+1} + c, \text{ for } n \neq 1$$

$$\int f'(x)\big[f(x)\big]^n dx = \frac{1}{n+1}\big[f(x)\big]^{n+1} + c, \text{ for } n \neq 1$$

Example 2

a $\int \sqrt{4x - 3}\,dx = \dfrac{(4x-3)^{\frac{3}{2}}}{6} + c$ **b** $\int x^2\sqrt{x^3 - 1}\,dx = \dfrac{2}{9}(x^3 - 1)^{\frac{3}{2}} + c$

Integrals involving trigonometric functions

$$\int \sin(ax + b)\,dx = -\frac{1}{a}\cos(ax + b) + c$$

$$\int \cos(ax + b)\,dx = \frac{1}{a}\sin(ax + b) + c$$

$$\int \sec^2(ax + b)\,dx = \frac{1}{a}\tan(ax + b) + c$$

$$\int f'(x)\sin f(x)\,dx = -\cos f(x) + c$$

$$\int f'(x)\cos f(x)\,dx = \sin f(x) + c$$

$$\int f'(x)\sec^2 f(x)\,dx = \tan f(x) + c$$

> **Hint**
> The $f(x)$ forms of the 3 bottom integrals appear on the HSC exam reference sheet.

9780170459228

Example 3

a $\displaystyle\int 2\sin 4x\,dx = -\frac{1}{2}\cos 4x + c$

b $\displaystyle\int 3x\sec^2(3x^2)\,dx = \frac{1}{2}\tan(3x^2) + c$

Integrals involving exponential functions

$$\int e^x\,dx = e^x + c$$

$$\int e^{ax+b}\,dx = \frac{1}{a}e^{ax+b} + c$$

$$\int a^x\,dx = \frac{1}{\ln a}a^x + c$$

$$\int f'(x)e^{f(x)}\,dx = e^{f(x)} + c$$

$$\int f'(x)a^{f(x)}\,dx = \frac{a^{f(x)}}{\ln a} + c$$

> **Hint**
>
> The $f(x)$ forms of integrals appear on the HSC exam reference sheet. Before the exam, know which integrals are provided and which ones you need to remember.

Example 4

a $\displaystyle\int 3^x\,dx = \frac{1}{\ln 3}3^x + c$

b $\displaystyle\int e^{3-x}\,dx = -e^{3-x} + c$

Integrals involving logarithmic functions

$$\int \frac{1}{x}\,dx = \ln|x| + c$$

$$\int \frac{f'(x)}{f(x)}\,dx = \ln|f(x)| + c$$

Example 5

a $\displaystyle\int \frac{3}{x-1}\,dx = 3\ln|x-1| + c$

b $\displaystyle\int \frac{x+1}{x^2+2x+4}\,dx = \frac{1}{2}\ln|x^2+2x+4| + c$

Applications of anti-differentiation

Given $f'(x)$ and an initial condition $f(a) = b$, find $f(x)$.

Example 6

The gradient function of a curve $y = f(x)$ is given by $f'(x) = 4x - 5$. The curve passes through the point $(3, 4)$. Determine the equation of the curve.

Solution

$f(x) = 2x^2 - 5x + c$

Passes through $(3, 4)$, so $f(3) = 4$.

$$4 = 2 \times 3^2 - 5 \times 3 + c$$
$$= 3 + c$$
$$c = 1$$

So $f(x) = 2x^2 - 5x + 1$.

9780170459228

C4.2 Areas and the definite integral

Approximating areas under a curve

We can use areas of squares, rectangles, triangles and trapeziums to approximate the area under a curve.

For example, if using rectangles, the approximation becomes more accurate as the number of rectangles increases.

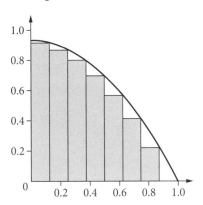 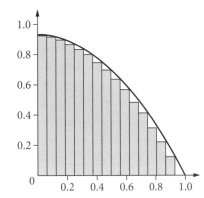 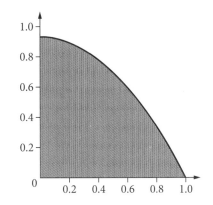

TOPIC SUMMARY

The trapezoidal rule

The **trapezoidal rule** uses the areas of trapeziums to approximate the area under a curve.

One subinterval or application of the trapezoidal rule (2 function values)

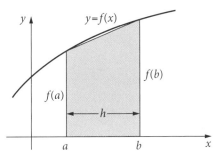

$A \approx \dfrac{h}{2}\left[f(a) + f(b)\right]$ using the formula for the area of a trapezium, where $h = b - a$.

Two subintervals or applications of the trapezoidal rule (3 function values)

Example 7

Use the trapezoidal rule to approximate the shaded area under the curve.

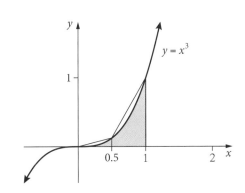

Solution

Apply $A \approx \dfrac{h}{2}\left[f(a) + f(b)\right]$ twice:

$$A \approx \frac{0.5}{2}\left[f(0) + f(0.5)\right] + \frac{0.5}{2}\left[f(0.5) + f(1)\right]$$

$$= \frac{0.5}{2}\left[f(0) + f(0.5) + f(0.5) + f(1)\right]$$

$$= 0.25\left[f(0) + 2f(0.5) + f(1)\right]$$

$$= 0.25\left[0^3 + 2(0.5^3) + 1^3\right]$$

$$= 0.3125 \text{ units}^2$$

Hint

Note that the middle side $f(0.5)$ is common to both trapeziums and needs to be counted twice.

n subintervals or applications of the rule (n +1 function values)

$$\int_a^b f(x)\,dx \approx \frac{h}{2}\{f(a) + f(b) + 2[f(x_1) + \cdots + f(x_{n-1})]\}, \text{ where } h = \frac{b-a}{n}, a = x_0, b = x_n$$

OR

$$\int_a^b f(x)\,dx \approx \frac{b-a}{2n}\{f(a) + f(b) + 2[f(x_1) + \cdots + f(x_{n-1})]\}$$

> **Hint**
> This 2nd formula is found on the HSC exam reference sheet.

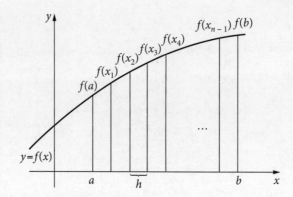

Note that, except for the first and last function values $f(a)$ and $f(b)$, the middle function values $f(x_1)$ to $f(x_{n-1})$ are common to 2 trapeziums and need to be counted twice.

Example 8

Use the trapezoidal rule to approximate $\int_2^5 \frac{1}{x-1}\,dx$ using 4 subintervals. Give the answer correct to two decimal places.

Solution

There are 4 subintervals, so 5 function values and $h = \dfrac{5-2}{4} = \dfrac{3}{4} = 0.75$.

x	2	2.75	3.5	4.25	5
$f(x)$	1	0.5714	0.4	0.3077	0.25

$$\int_2^5 \frac{1}{x-1}\,dx \approx \frac{h}{2}\{f(a) + f(b) + 2[f(x_1) + f(x_2) + f(x_3)]\}$$

$$= \frac{0.75}{2}\{1 + 0.25 + 2[0.5714 + 0.4 + 0.3077]\}$$

$$\approx 1.43 \text{ units}^2$$

The definite integral $\int_a^b f(x)\,dx$

$\int_a^b f(x)\,dx$ calculates the area under the curve of $y = f(x)$ from $x = a$ to $x = b$, which is shown by the shaded area *above* the x-axis.

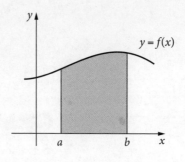

Fundamental theorem of calculus

The area enclosed by the curve $y = f(x)$, the x-axis and the lines $x = a$ and $x = b$ is given by the definite integral $\int_a^b f(x)\,dx = F(b) - F(a)$, where $F(x)$ is the anti-derivative of $f(x)$.

Example 9

$$\int_{-2}^{2} x^2\,dx = \left[\frac{x^3}{3}\right]_{-2}^{2}$$

$$= \frac{2^3}{3} - \frac{(-2)^3}{3}$$

$$= 5\frac{1}{3}\ \text{units}^2$$

Hint
With definite integrals, the answer is a number.

Example 10

a Differentiate $y = xe^{4x}$.

b Hence, find the exact value of $\int_0^3 e^{4x}(16x + 4)\,dx$.

Solution

a Using the product rule with:

$u = x \qquad v = e^{4x}$
$u' = 1 \qquad v' = 4e^{4x}$

$y' = x\,4e^{4x} + e^{4x}(1)$
$\quad = e^{4x}(4x + 1)$

b $\int_0^3 e^{4x}(16x + 4)\,dx = 4\int_0^3 e^{4x}(4x + 1)\,dx$

$$= 4\left[xe^{4x}\right]_0^3$$

$$= 4\left(3e^{4\times 3} - 0\right)$$

$$= 12e^{12}$$

Areas under curves

Example 11

Use the trapezoidal rule to approximate the shaded area under the curve.

Solution

Area of shaded region $= \int_1^2 \frac{1}{x}\,dx$

$$= \left[\ln|x|\right]_1^2$$

$$= (\ln 2)\ \text{units}^2$$

Hint
Don't forget 'units^2'.

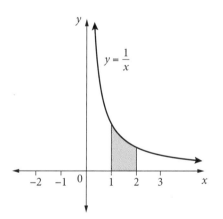

Areas **below** the x-axis give a **negative** definite integral, so $A = \left|\int_a^b f(x)\,dx\right|$ for the area enclosed by the curve $y = f(x)$ below the x-axis from $x = a$ to $x = b$.

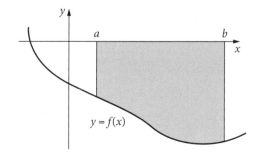

Example 12

Find the shaded area.

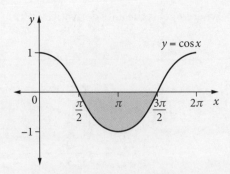

Solution

Area of shaded region $= \left| \int_{\frac{\pi}{2}}^{\frac{3\pi}{2}} \cos x \, dx \right| = \left| \left[\sin x \right]_{\frac{\pi}{2}}^{\frac{3\pi}{2}} \right| = \left| -2 \right| = 2 \text{ units}^2$

If the area has different sections above and below the x-axis, then each separate section must be calculated by integration and added together.

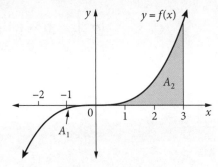

$A_1 = \left| \int_{-1}^{0} f(x) \, dx \right|$ for the area under the curve $y = f(x)$ from $x = -1$ to $x = 0$, which shows the shaded area below the x-axis.

$A_2 = \int_{0}^{3} f(x) \, dx$ for the area under the curve $y = f(x)$ from $x = 0$ to $x = 3$, which shows the shaded area **above** the x-axis.

Area of shaded region $= A_1 + A_2 = \left| \int_{-1}^{0} f(x) \, dx \right| + \int_{0}^{3} f(x) \, dx$

Even and odd functions

Even function

$f(-x) = f(x)$

Has line symmetry about the y-axis

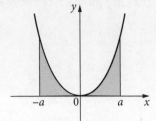

Equal areas,
both positive.

$\int_{-a}^{a} f(x) \, dx = 2 \int_{0}^{a} f(x) \, dx$

Odd function

$f(-x) = -f(x)$

Has point symmetry about the origin

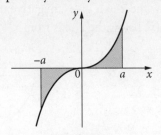

Equal areas,
but opposite signs.

$\int_{-a}^{a} f(x) \, dx = 0$

9780170459228

Areas between curves

$$A = \int_a^b \left[f(x) - g(x) \right] dx$$

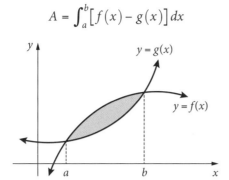

Example 13

Find the area enclosed between the curve of $y = x^2$ and $y = x + 2$.

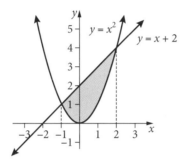

Solution

Area of shaded region $= \int_{-1}^{2} (x + 2) - x^2 \, dx$

$$= 4.5 \text{ units}^2$$

Applications of integration

Displacement is the integral of velocity; velocity is the integral of acceleration.

Problems involving motion and rates of change.

Example 14

A particle is moving in a straight line so that its velocity v m/s is given by $v = 4 - 8\cos 2t$, where t is the time, in seconds, after passing a fixed point O. Find in exact form:

a the time when the particle is at rest for the first time after starting its journey.

b the distance of the particle from point O at this time.

Answer

a $t = \dfrac{\pi}{6}$ seconds

b $\left(\dfrac{2\pi}{3} - 2\sqrt{3} \right)$m

Practice set 1

Multiple-choice questions

Solutions start on page 97.

Question 1 ○●●

Find $\int x^2 - 2x + 3\,dx$.

A $\dfrac{x^3}{3} - x^2 + c$　　**B** $\dfrac{x^3}{3} - x^2 + 3x + c$　　**C** $x^3 - 2x^2 + c$　　**D** $x^3 - 4x^2 + 3x + c$

Question 2 ○●●

Find $\int e^{3x}\,dx$.

A $e^{3x} + c$　　**B** $3e^{3x} + c$　　**C** $\dfrac{e^{3x}}{3} + c$　　**D** $\dfrac{e^{3x+1}}{3x+1} + c$

Question 3 ○●●

Find the anti-derivative of $\sec^2 x$.

A $\cos^2 x + c$　　**B** $x + \tan^2 x + c$　　**C** $\tan x + c$　　**D** $\operatorname{cosec}^2 x + c$

Question 4 ○●●

Evaluate $\int_2^5 \dfrac{1}{y^3}\,dy$.

A -0.58　　**B** 0.0203　　**C** 0.105　　**D** 0.42

Question 5 ○●●

Find $\int 2y^{-\frac{1}{2}}\,dy$.

A $4\sqrt{y} + c$　　**B** $\sqrt{y} + c$　　**C** $2\sqrt{y^3} + c$　　**D** $\dfrac{4}{3\sqrt{y^3}} + c$

Question 6 ○●●

Evaluate $\int_0^{\frac{\pi}{6}} 3\cos 4x\,dx$.

A $\dfrac{3}{8}$　　**B** $\dfrac{3\sqrt{3}}{8}$　　**C** 6　　**D** $6\sqrt{3}$

Question 7 ○●●

Evaluate $\int_1^2 e^x + 3\,dx$.

A $e^2 - e + 3$　　**B** $e^2 - e$　　**C** $e^2 - e + 9$　　**D** $e^2 + e - 3$

Question 8 ○●●

Evaluate $\int_e^{e^3} \dfrac{1}{5x}\,dx$.

A $\dfrac{2}{5}$　　**B** $\dfrac{3}{5}$　　**C** 2　　**D** 10

Question 9 ⦿⦿⦾

Evaluate $\int_1^4 \dfrac{2x+5}{x}\,dx$.

A $6+5\ln 2$ **B** $6+10\ln 2$ **C** $5\ln 2$ **D** $10\ln 2$

Question 10 ⦿⦿⦾

$y = H(x)$ is an odd function where $\int_0^a H(x)\,dx = 7$.

Find the value of $\int_{-a}^a H(x)\,dx$.

A 0 **B** 7

C 14 **D** not possible to determine

Question 11 ⦿⦿⦾

Using the trapezoidal rule with 5 function values, which expression gives the approximate area under

the curve $y = \dfrac{x}{e^x}$ from $x = 2$ to $x = 4$?

A $\dfrac{1}{4}\left[\dfrac{2}{e^2} + \dfrac{5}{e^{2.5}} + \dfrac{6}{e^3} + \dfrac{7}{e^{3.5}} + \dfrac{4}{e^4}\right]$ **B** $\dfrac{1}{2}\left[\dfrac{1}{e^2} + \dfrac{10}{e^{2.5}} + \dfrac{12}{e^3} + \dfrac{14}{e^{3.5}} + \dfrac{2}{e^4}\right]$

C $\dfrac{1}{2}\left[\dfrac{1}{e^2} + \dfrac{5}{e^{2.5}} + \dfrac{6}{e^3} + \dfrac{7}{e^{3.5}} + \dfrac{2}{e^4}\right]$ **D** $\dfrac{1}{4}\left[\dfrac{2}{e^2} + \dfrac{5}{2e^{2.5}} + \dfrac{3}{e^3} + \dfrac{7}{2e^{3.5}} + \dfrac{4}{e^4}\right]$

Question 12 ⦿⦿⦾

The graph of $y = f(x)$ is shown below from $x = -6$ to $x = 4$ with A_1, A_2 and A_3 representing each shaded area.

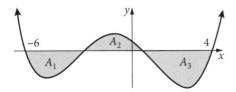

$\int_{-6}^4 f(x)\,dx = 8$ and $A_1 + A_2 + A_3 = 26$.

What is the value of A_2?

A 9 **B** 12 **C** 17 **D** 18

Question 13 ⦿⦿⦾

The diagram shows the parabola $y = 6x - x^2$ intersecting with the line $y = 3x$.

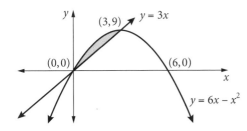

Which expression shows the integral representing the area of the shaded region, bounded by the parabola and the line?

A $\int_0^3 x^2 - 3x\,dx$ **B** $\int_0^3 3x - x^2\,dx$ **C** $\int_0^6 x^2 - 3x\,dx$ **D** $\int_0^6 3x - x^2\,dx$

Question 14 ●●

Find the value of $\int_{-4}^{1} |x + 2| \, dx$.

A $\dfrac{5}{2}$ 　　　　　　 **B** $\dfrac{13}{2}$ 　　　　　　 **C** $\dfrac{19}{2}$ 　　　　　　 **D** $\dfrac{37}{2}$

Question 15 ●●

The area bounded by the line $y = 5$ and the curve $y = x^2 - 4$ is shown in the diagram.

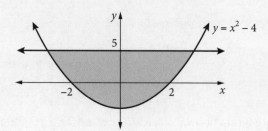

Which integral represents the shaded area?

A $\int_{-2}^{2} 9 - x^2 \, dx$ 　　　　　　　　　　　 **B** $2\int_{0}^{3} 9 - x^2 \, dx$

C $\int_{-3}^{3} 1 - x^2 \, dx$ 　　　　　　　　　　　 **D** $2\int_{0}^{2} 1 - x^2 \, dx$

Question 16 ●●●

Which expression below can be used to find the area of the shaded region shown in the diagram?

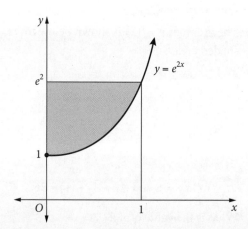

A $\int_{0}^{1} e^{2x} \, dx$

B $e^2 - \int_{0}^{1} e^{2x} \, dx$

C $\int_{1}^{e^2} e^{2y} \, dy$

D $\int_{1}^{e^2} e^{2x} - 1 \, dx$

Question 17 ●●●

Evaluate $\int_{0}^{1} \dfrac{e}{e^{2x}} \, dx$.

A $\dfrac{e^3 - 3}{3e^2}$ 　　　　　　　　　　　 **B** $\dfrac{e^3 - 1}{3e^2}$

C $\dfrac{e^3 - 3}{e}$ 　　　　　　　　　　　 **D** $\dfrac{e^2 - 1}{2e}$

Question 18 ●●●

Simplify $\int_{0}^{\frac{\pi}{3}} m\cos x - n\sin x \, dx$.

A $\dfrac{1}{2}(m + \sqrt{3}n)$ 　　　　　　　　 **B** $\dfrac{1}{2}(n - \sqrt{3}m)$

C $-\dfrac{1}{2}(m + \sqrt{3}n)$ 　　　　　　　 **D** $\dfrac{1}{2}(\sqrt{3}m - n)$

Question 19 ●●●

Rodrigo approximates the area under the graph of $y = 2\sin 2x + 4$ between $x = 0$ and $x = \dfrac{\pi}{2}$ using the 3 rectangles shown.

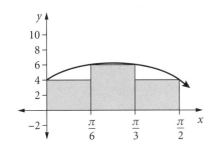

What is the difference between the exact area under the graph and Rodrigo's approximated area?

A $2 - \dfrac{\sqrt{3}\pi}{6}$

B $2 + \dfrac{\sqrt{3}\pi}{3} - 2\pi$

C $4\pi - \dfrac{\sqrt{3}\pi}{6} + 2$

D $4 - \dfrac{\sqrt{3}\pi}{3}$

Question 20 ●●●

The graph of $y = f'(x)$ is shown.

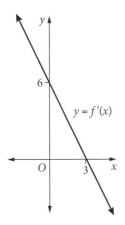

What is the equation of y if it has a maximum value of 16?

A $y = x^2 - 6x + 16$

B $y = x^2 - 6x - 11$

C $y = 7 + 6x - x^2$

D $y = 25 + 6x - x^2$

Practice set 2

Short-answer questions

Solutions start on page 100.

Question 1 (5 marks) ⬤○○
Find each integral.

a $\int 2x - 1 \, dx$ 1 mark

b $\int e^{2x} \, dx$ 1 mark

c $\int \cos 3x \, dx$ 1 mark

d $\int \frac{4}{x} \, dx$ 1 mark

e $\int (5x - 1)^4 \, dx$ 1 mark

Question 2 (2 marks) ⬤⬤○
Find $\int (x + 1)^2 \, dx$.   2 marks

Question 3 (6 marks) ⬤⬤○
Evaluate each definite integral.

a $\int_0^1 \sin \pi x \, dx$ 2 marks

b $\int_0^\pi \cos \frac{x}{4} \, dx$ 2 marks

c $\int_0^{\frac{\pi}{6}} \sec^2 2x \, dx$ 2 marks

Question 4 (2 marks) ©NESA 2019 HSC EXAM, QUESTION 14(b)(ii) ⬤○○
The derivative of a function $y = f(x)$ is given by $f'(x) = 3x^2 + 2x - 1$.

The curve passes through the point $(0, 4)$. Find an expression for $f(x)$.   2 marks

Question 5 (10 marks) ⬤⬤○
Find each integral.

a $\int \frac{x^6 - 2x}{x^2} \, dx$ 2 marks

b $\int x + \sin \frac{x}{2} \, dx$ 2 marks

c $\int \frac{x - 1}{x^2 - 2x + 3} \, dx$ 2 marks

d $\int \sqrt{x}(1 - x) \, dx$ 2 marks

e $\int \sin x° \, dx$ 2 marks

Question 6 (4 marks)

Evaluate each integral in exact form.

a $\int_0^{\ln 5} e^{2x}\, dx$ 2 marks

b $\int_1^2 3^x\, dx$ 2 marks

Question 7 (2 marks)

Given $\int_0^3 f(x)\, dx = 4$, find $\int_0^3 f(x) + 3\, dx$.   2 marks

Question 8 (2 marks)

Given that $\int_0^4 (x^2 + k)\, dx = 10$, find the value of k.   2 marks

Question 9 (3 marks) ©NESA 2020 HSC EXAM, QUESTION 18

a Differentiate $e^{2x}(2x + 1)$.   2 marks

b Hence, find $\int (x + 1)e^{2x}\, dx$.   1 mark

Question 10 (3 marks)

This diagram shows the graph of $y = \dfrac{5x}{x^2 + 1}$.

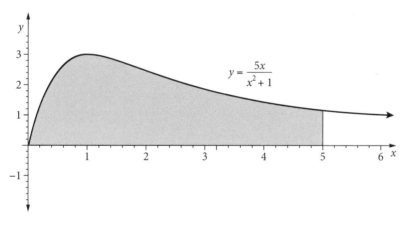

$y = \dfrac{5x}{x^2 + 1}$

Calculate the exact value of the shaded region.   3 marks

Question 11 (2 marks)

This diagram shows a cross-section of a stormwater channel of width 12 metres with vertical banks.

The depth of the water, in metres, is measured at 3-metre intervals across its cross-section and the measurements recorded as shown in the table below.

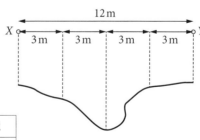

Distance from X (m)	0	3	6	9	12
Depth (m)	0.95	1.25	1.68	1.17	0.92

Calculate the area of the cross-section using the trapezoidal rule with 5 function values.   2 marks

PRACTICE SET 2

Question 12 (2 marks)

The shaded region is enclosed by the curve $y = x^3 - 3x$ and the line $y = x$, as shown in the diagram below. The line $y = x$ meets the curve $y = x^3 - 3x$ at $O(0, 0)$ and $A(2, 2)$.

Find the area of the shaded region.   2 marks

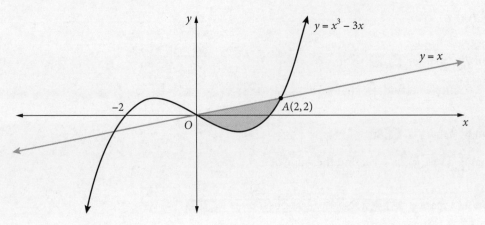

Question 13 (5 marks)

a   Differentiate $y = x \sin x$.   2 marks

b   Hence or otherwise, find the exact value of $\int_0^{\frac{\pi}{6}} x \cos x \, dx$.   3 marks

Question 14 (4 marks)

The diagram below shows the function $y = h(x)$ whose graph consists of a semicircle of radius 3 and a straight line.

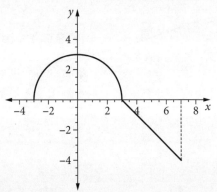

a   Calculate the value of $\int_{-3}^{7} h(x) \, dx$.   2 marks

b   Find the area contained between the graph of $y = h(x)$ and the x-axis.   1 mark

c   Explain why the value of the integral in part **a** is not the same as the value of the area in part **b**.   1 mark

9780170459228

Question 15 (5 marks) ●●●

The curves $y = \sin x$ and $y = \cos x$ are shown below for $[0, 2\pi]$.

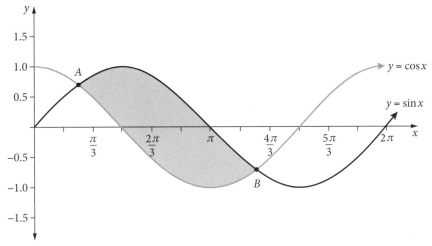

a Show that the x-coordinates of the points of intersection A and B are $\dfrac{\pi}{4}$ and $\dfrac{5\pi}{4}$ respectively.   2 marks

b Hence, find the shaded area.   3 marks

Question 16 (3 marks) ●●

Find the area enclosed by the curves $y = (x + 1)^2$ and $y = (x - 4)^2$ and the x-axis, as shown in the diagram below.   3 marks

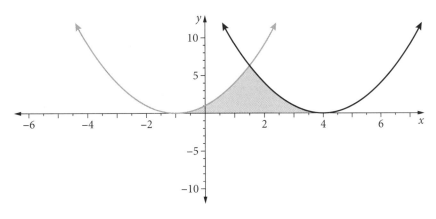

Question 17 (4 marks) ●●●

The graphs of $y = \sin\dfrac{\pi x}{2}$ and $y = \cos \pi x$ are shown in the diagram below.

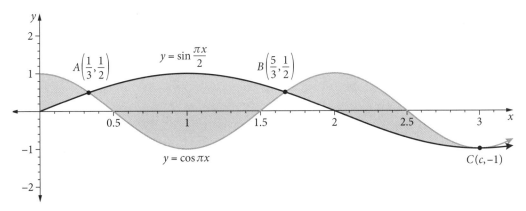

a The graphs intersect at 3 points, $A\left(\dfrac{1}{3}, \dfrac{1}{2}\right)$, $B\left(\dfrac{5}{3}, \dfrac{1}{2}\right)$ and $C(c, -1)$ for $[0, \pi]$. Find the value of c.   2 marks

b Calculate the total area of the 3 shaded regions.   2 marks

Question 18 (6 marks) ●●●

The diagram below shows the graph of $y = \frac{1}{2}\log_e x$.

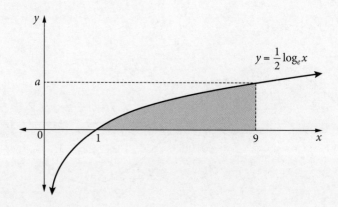

a   Make x the subject of the equation $y = \frac{1}{2}\log_e x$.   1 mark

b   Find the value of a such that $f(9) = a$.   2 marks

c   Calculate the shaded area, leaving your answer in exact form.   3 marks

Question 19 (3 marks)   ©NESA   2017 HSC EXAM, QUESTION 13(d)   ●●○

The rate at which water flows into a tank is given by $\dfrac{dV}{dt} = \dfrac{2t}{1+t^2}$, where V is the volume
of water in the tank in litres and t is the time in seconds.

Initially the tank is empty.

Find the exact amount of water in the tank after 10 seconds.   3 marks

Question 20 (9 marks) ●●●

A motorcyclist travels along a straight road so that t seconds after leaving his starting point, his
velocity, v m/s, is given by $v = 10t - t^2$. The motorcyclist reached a maximum speed at $t = 5$ and
maintained this speed for another 5 seconds, before coming to rest with a constant acceleration
of $a = -4 \, \text{m/s}^2$.

a   Find the constant speed during the second 5 seconds.   1 mark

b   Find the equation for the velocity, in terms of t, during the period of constant acceleration.   2 marks

c   For the whole of the motion, sketch the velocity–time graph.   3 marks

d   Calculate, correct to the nearest metre, the total distance travelled.   3 marks

Practice set 1

Worked solutions

1 B

$$\int x^2 - 2x + 3\, dx = \frac{x^3}{3} - x^2 + 3x + c$$

2 C

$$\int e^{3x}\, dx = \frac{1}{3}e^{3x} + c$$

3 C

$$\int \sec^2 x\, dx = \tan x + c$$

4 C

$$\int_2^5 \frac{1}{y^3}\, dy = \int_2^5 y^{-3}\, dy$$

$$= \left[\frac{y^{-2}}{-2}\right]_2^5$$

$$= \left[\frac{-1}{2y^2}\right]_2^5$$

$$= \frac{-1}{2 \times 5^2} + \frac{1}{2 \times 2^2}$$

$$= \frac{-1}{50} + \frac{1}{8}$$

$$= \frac{21}{200}$$

$$= 0.105$$

5 A

$$\int 2y^{-\frac{1}{2}}\, dy = 2 \times 2y^{\frac{1}{2}} + c$$

$$= 4\sqrt{y} + c$$

6 B

$$\int_0^{\frac{\pi}{6}} 3\cos 4x\, dx = \frac{3}{4}[\sin 4x]_0^{\frac{\pi}{6}}$$

$$= \frac{3}{4}\left(\sin\frac{2\pi}{3} - 0\right)$$

$$= \frac{3}{4} \times \frac{\sqrt{3}}{2}$$

$$= \frac{3\sqrt{3}}{8}$$

7 A

$$\int_1^2 e^x + 3\, dx = \left[e^x + 3x\right]_1^2$$

$$= e^2 + 3 \times 2 - (e + 3)$$

$$= e^2 + 6 - e - 3$$

$$= e^2 - e + 3$$

8 A

$$\frac{1}{5}\int_e^{e^3} \frac{1}{x}\, dx = \frac{1}{5}\left[\ln x\right]_e^{e^3}$$

$$= \frac{1}{5}\ln e^3 - \frac{1}{5}\ln e$$

$$= \frac{2}{5}\ln e$$

$$= \frac{2}{5} \times 1$$

$$= \frac{2}{5}$$

9 B

$$\int_1^4 \frac{2x + 5}{x}\, dx = \int_1^4 2 + \frac{5}{x}\, dx$$

$$= \left[2x + 5\ln|x|\right]_1^4$$

$$= 2 \times 4 + 5\ln 4 - (2 + 5\ln 1$$

$$= 8 + 5\ln 4 - 2 - 5 \times 0$$

$$= 6 + 5\ln 4$$

$$= 6 + 5\ln 2^2$$

$$= 6 + 10\ln 2$$

10 A

$y = H(x)$ is an odd function.

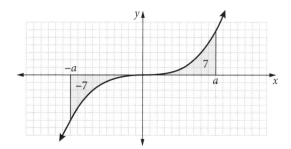

The graph shows an example of an odd function. It has point symmetry about the origin.

$$\int_{-a}^0 H(x)dx = -7 \text{ and } \int_0^a H(x)dx = 7$$

So $\int_{-a}^a H(x)dx = 7 - 7 = 0$.

11 A

$$\int_2^4 \frac{x}{e^x}\,dx \approx \frac{\left(\frac{4-2}{4}\right)}{2}\big[f(2) + 2[f(2.5) + f(3) + f(3.5)] + f(4)\big]$$

$$= \frac{1}{4}\left[\frac{2}{e^2} + 2\left[\frac{2.5}{e^{2.5}} + \frac{3}{e^3} + \frac{3.5}{e^{3.5}}\right] + \frac{4}{e^4}\right]$$

$$= \frac{1}{4}\left[\frac{2}{e^2} + \frac{5}{e^{2.5}} + \frac{6}{e^3} + \frac{7}{e^{3.5}} + \frac{4}{e^4}\right]$$

12 C

$$A_1 + A_2 + A_3 = 26 \qquad [1]$$

$$\int_{-6}^4 f(x)\,dx = 8$$

$$-A_1 + A_2 - A_3 = 8 \qquad [2]$$

$$[1] + [2]$$

$$2A_2 = 26 + 8$$
$$ = 34$$
$$A_2 = 17 \text{ units}^2$$

13 B

$$A = \int_0^3 6x - x^2 - 3x\,dx$$

$$= \int_0^3 3x - x^2\,dx$$

14 B

For $\int_{-4}^1 |x + 2|\,dx$

The absolute value graph from $x = -4$ to $x = 1$ creates 2 triangles above the x-axis.

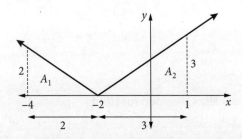

$$A_1 = \tfrac{1}{2} \times 2 \times 2 \qquad\qquad A_2 = \tfrac{1}{2} \times 3 \times 3$$
$$ = 2 \qquad\qquad\qquad\qquad = 4.5$$

$$\int_{-4}^1 |x + 2|\,dx = 2 + 4.5 = 6.5$$

15 B

$$2\int_0^3 5 - (x^2 - 4)\,dx = 2\int_0^3 9 - x^2\,dx$$

16 B

Shaded area = Area of rectangle – area under curve (or area between horizontal line $y = e^2$ and $y = e^{2x}$)

$$\text{Area} = 1 \times e^2 - \int_0^1 e^{2x}\,dx \text{ OR } \int_0^1 e^2 - e^{2x}\,dx$$

$$= e^2 - \int_0^1 e^{2x}\,dx$$

17 D

$$\int_0^1 e \times e^{-2x}\,dx = [e^{1-2x}]_0^1$$

$$= \left[-\frac{1}{2}e^{1-2x}\right]_0^1$$

$$= -\frac{1}{2}[e^{-1} - e]$$

$$= -\frac{1}{2}\left[\frac{1}{e} - e\right]$$

$$= -\frac{1}{2}\left(\frac{1 - e^2}{e}\right)$$

$$= \frac{e^2 - 1}{2e}$$

18 D

$$\int_0^{\frac{\pi}{3}} m\cos x - n\sin x \, dx = \left[m\sin x + n\cos x \right]_0^{\frac{\pi}{3}}$$

$$= m\sin\left(\frac{\pi}{3}\right) + n\cos\left(\frac{\pi}{3}\right) - (m\sin 0 + n\cos 0)$$

$$= m \times \frac{\sqrt{3}}{2} + n \times \frac{1}{2} - n$$

$$= \frac{\sqrt{3}m}{2} + \frac{n}{2} - n$$

$$= \frac{\sqrt{3}m}{2} - \frac{n}{2}$$

$$= \frac{1}{2}(\sqrt{3}m - n)$$

19 A

$$A = \int_0^{\frac{\pi}{2}} 2\sin 2x + 4 \, dx = \left[-\cos 2x + 4x \right]_0^{\frac{\pi}{2}}$$

$$= -\cos\pi + \frac{4\pi}{2} - (-\cos 0 + 0)$$

$$= -(-1) + 2\pi + 1$$

$$= (2\pi + 2) \text{ units}^2$$

Using the 3 rectangles:

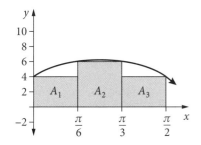

$$A_1 = A_3 = \frac{\pi}{6} \times 4$$

$$= \frac{2\pi}{3}$$

$$A_2 = \frac{\pi}{6} \times (\sqrt{3} + 4)$$

$$= \frac{\sqrt{3}\pi}{6} + \frac{2\pi}{3}$$

Total area $= \frac{2\pi}{3} \times 3 + \frac{\sqrt{3}\pi}{6}$

$$= \frac{\sqrt{3}\pi}{6} + 2\pi \text{ units}^2$$

Difference $= 2\pi + 2 - \left(2\pi + \frac{\sqrt{3}\pi}{6} \right)$

$$= 2\pi + 2 - 2\pi - \frac{\sqrt{3}\pi}{6}$$

$$= 2 - \frac{\sqrt{3}\pi}{6} \text{ units}^2$$

20 C

$$y' = -2x + 6$$

$$y = \frac{-2x^2}{2} + 6x + c$$

$$y' = 0$$

$$0 = -2x + 6$$

$$2x = 6$$

$$x = 3$$

$$y = -3^2 + 6 \times 3 + c = 16$$

$$-9 + 18 + c = 16$$

$$9 + c = 16$$

$$c = 7$$

So $y = -x^2 + 6x + 7$.

Practice set 2

Worked solutions

Question 1

a $\int 2x - 1\, dx = \dfrac{2x^2}{2} - x + c$

$\quad = x^2 - x + c$

b $\int e^{2x}\, dx = \dfrac{1}{2}e^{2x} + c$

c $\int \cos 3x\, dx = \dfrac{1}{3}\sin 3x + c$

d $\int \dfrac{4}{x}\, dx = 4\ln|x| + c$

e $\int (5x - 1)^4\, dx = \dfrac{(5x - 1)^5}{5 \times 5} + c$

$\quad = \dfrac{(5x - 1)^5}{25} + c$

Question 2

$\int (x + 1)^2\, dx = \int x^2 + 2x + 1\, dx$

$\quad = \dfrac{x^3}{3} + x^2 + x + c$

OR

$\int (x + 1)^2\, dx = \dfrac{(x + 1)^3}{3} + c$

Question 3

a $\int_0^1 \sin \pi x\, dx = \dfrac{-1}{\pi}\left[\cos \pi - \cos 0\right]$

$\quad = \dfrac{-1}{\pi}(-1 - 1)$

$\quad = \dfrac{-1}{\pi}(-2)$

$\quad = \dfrac{2}{\pi}$

b $\int_0^\pi \cos \dfrac{x}{4}\, dx = \left[4\sin \dfrac{\pi}{4}\right]_0^\pi$

$\quad = 4\sin \dfrac{\pi}{4} - 4\sin 0$

$\quad = 4 \times \dfrac{1}{\sqrt{2}}$

$\quad = \dfrac{4\sqrt{2}}{2}$

$\quad = 2\sqrt{2}$

c $\int_0^{\frac{\pi}{6}} \sec^2 2x\, dx = \dfrac{1}{2}\left[\tan 2x\right]_0^{\frac{\pi}{6}}$

$\quad = \dfrac{1}{2}\left(\tan \dfrac{\pi}{3} - \tan 0\right)$

$\quad = \dfrac{1}{2}(\sqrt{3} - 0)$

$\quad = \dfrac{\sqrt{3}}{2}$

Question 4

$f'(x) = 3x^2 + 2x - 1$

$f(x) = x^3 + x^2 - x + c$

When $x = 0, f(x) = 4$.

$4 = 0^3 + 0^2 - 0 + c$

$c = 4$

So $f(x) = x^3 + x^2 - x + 4$.

Question 5

a $\int \dfrac{x^6 - 2x}{x^2}\, dx = \int \dfrac{x^6}{x^2} - \dfrac{2x}{x^2}\, dx$

$\quad = \int x^4 - \dfrac{2}{x}\, dx$

$\quad = \dfrac{x^5}{5} - 2\ln|x| + c$

b $\int x + \sin \dfrac{x}{2}\, dx = \dfrac{x^2}{2} - 2\cos \dfrac{x}{2} + c$

c For $f(x) = x^2 - 2x + 3$

and $f'(x) = 2x - 2$

$\int \dfrac{x - 1}{x^2 - 2x + 3}\, dx = \int \dfrac{\frac{1}{2}(2x - 2)}{x^2 - 2x + 3}\, dx$

$\quad = \dfrac{1}{2}\ln\left|x^2 - 2x + 3\right| + c$

d $\int \sqrt{x}(1 - x)\, dx = \int \sqrt{x} - x\sqrt{x}\, dx$

$\quad = \int x^{\frac{1}{2}} - x^{\frac{3}{2}}\, dx$

$\quad = \dfrac{2}{3}x^{\frac{3}{2}} - \dfrac{2}{5}x^{\frac{5}{2}} + c$

$\quad = \dfrac{2}{3}\sqrt{x^3} - \dfrac{2}{5}\sqrt{x^5} + c$

e $\int \sin x^\circ\, dx = \int \sin\left(\dfrac{\pi x}{180}\right)\, dx$

$\quad = -\dfrac{180}{\pi}\cos\left(\dfrac{\pi x}{180}\right) + c$

> **Hint**
> Convert degrees to radians before integrating.

Question 6

a $\int_0^{\ln 5} e^{2x}\,dx = \left[\frac{1}{2}e^{2x}\right]_0^{\ln 5}$

$$= \frac{1}{2}\left(e^{2\ln 5} - e^0\right)$$

$$= \frac{1}{2}\left(e^{\ln 5^2} - 1\right)$$

$$= \frac{1}{2}(25 - 1)$$

$$= \frac{1}{2} \times 24$$

$$= 12$$

b $\int_1^2 3^x\,dx = \frac{1}{\ln 3}\left[3^x\right]_1^2$

$$= \frac{1}{\ln 3}\left(3^2 - 3\right)$$

$$= \frac{1}{\ln 3}(9 - 3)$$

$$= \frac{6}{\ln 3}$$

Question 7

$$\int_0^3 f(x)\,dx = 4 \quad \text{(given)}$$

$$\int_0^3 f(x) + 3\,dx = \int_0^3 f(x)\,dx + \int_0^3 3\,dx$$

$$= 4 + [3x]_0^3$$

$$= 4 + (3 \times 3 - 0)$$

$$= 13$$

Question 8

$$\int_0^4 x^2 + k\,dx = 10$$

$$\left[\frac{x^3}{3} + kx\right]_0^4 = 10$$

$$\frac{4^3}{3} + 4k - 0 = 10$$

$$64 + 12k = 30$$

$$12k = -34$$

$$k = -\frac{17}{6}$$

Question 9

a $y = e^{2x}(2x + 1)$

By the product rule,

$$\frac{dy}{dx} = e^{2x}(2) + 2e^{2x}(2x + 1)$$

$$= 2e^{2x} + 4xe^{2x} + 2e^{2x}$$

$$= 4xe^{2x} + 4e^{2x}$$

$$= 4e^{2x}(x + 1)$$

b From part **a**,

$$\int 4e^{2x}(x + 1)\,dx = e^{2x}(2x + 1) + c$$

$$4\int e^{2x}(x + 1)\,dx = e^{2x}(2x + 1) + c$$

$$\int e^{2x}(x + 1)\,dx = \frac{1}{4}e^{2x}(2x + 1) + c$$

Question 10

$$A = \int_0^5 \frac{5x}{x^2 + 1}\,dx = \left[\frac{5}{2}\ln\left(x^2 + 1\right)\right]_0^5$$

$$= \frac{5}{2}\ln\left(5^2 + 1\right) - \frac{5}{2}\ln 1$$

$$= \frac{5}{2}\ln 26 - \frac{5}{2} \times 0$$

$$= \frac{5}{2}\ln 26 \text{ units}^2$$

Question 11

$$\text{Area} = \frac{3}{2}\Big[f(0) + 2\{f(3) + f(6) + f(9)\} + f(12)\Big]$$

$$= \frac{3}{2}\Big[0.95 + 2\{1.25 + 1.68 + 1.17\} + 0.92\Big]$$

$$= \frac{3}{2}(10.07)$$

$$= 15.105\,\text{m}^2$$

Question 12

$$A = \int_0^2 x - (x^3 - 3x)\,dx$$

$$= \int_0^2 4x - x^3\,dx$$

$$= \left[2x^2 - \frac{x^4}{4}\right]_0^2$$

$$= 2 \times 2^2 - \frac{16}{4} - 0$$

$$= 4 \text{ units}^2$$

Question 13

a $y = x \sin x$

$y' = x \cos x + \sin x$

b $\int_0^{\frac{\pi}{6}} x \cos x + \sin x \, dx = \left[x \sin x \right]_0^{\frac{\pi}{6}}$

Rearranging,

$\int_0^{\frac{\pi}{6}} x \cos x \, dx = \left[x \sin x \right]_0^{\frac{\pi}{6}} - \int_0^{\frac{\pi}{6}} \sin x \, dx$

$= \frac{\pi}{6} \sin \frac{\pi}{6} - 0 + \left[\cos x \right]_0^{\frac{\pi}{6}}$

$= \frac{\pi}{6} \times \frac{1}{2} + \cos \frac{\pi}{6} - \cos 0$

$= \frac{\pi}{12} + \frac{\sqrt{3}}{2} - 1 \text{ units}^2$

Question 14

a $\int_{-3}^{7} h(x) \, dx = \frac{1}{2} \times \pi \times 3^2 - \frac{1}{2} \times 4 \times 4$

$= \frac{9\pi}{2} - 8$

b $A = \frac{9\pi}{2} + |-8|$

$= \left(\frac{9\pi}{2} + 8 \right) \text{ units}^2$

c By integration, the semicircle is above the x-axis and has a positive value $\left(\frac{9\pi}{2} \right)$, whereas the triangle is below the x-axis and has a negative integral value. Area can only take a positive value so $\left| \int_{-3}^{7} h(x) \, dx \right| = 8$ and must be *added* to $\frac{9\pi}{2}$.

Question 15

a $\cos x = \sin x$

$1 = \frac{\sin x}{\cos x}$

$\tan x = 1$

$x = \frac{\pi}{4}, \pi + \frac{\pi}{4}$

$= \frac{\pi}{4}, \frac{5\pi}{4}$

$A: x = \frac{\pi}{4}$

$B: x = \frac{5\pi}{4}$

b $A = \int_{\frac{\pi}{4}}^{\frac{5\pi}{4}} \sin x - \cos x \, dx = \left[-\cos x - \sin x \right]_{\frac{\pi}{4}}^{\frac{5\pi}{4}}$

$= -\cos \frac{5\pi}{4} - \sin \frac{5\pi}{4} - \left(-\cos \frac{\pi}{4} - \sin \frac{\pi}{4} \right)$

$= -\left(-\frac{1}{\sqrt{2}} - \frac{1}{\sqrt{2}} \right) + \frac{1}{\sqrt{2}} + \frac{1}{\sqrt{2}}$

$= \frac{4}{\sqrt{2}}$

$= 2\sqrt{2} \text{ units}^2$

Question 16

$(x + 1)^2 = (x - 4)^2$

$x^2 + 2x + 1 = x^2 - 8x + 16$

$10x = 15$

$x = \frac{3}{2}$

$A = \int_{-1}^{\frac{3}{2}} (x + 1)^2 \, dx + \int_{\frac{3}{2}}^{4} (x - 4)^2 \, dx$

$= \left[\frac{(x + 1)^3}{3} \right]_{-1}^{\frac{3}{2}} + \left[\frac{(x - 4)^3}{3} \right]_{\frac{3}{2}}^{4}$

$= \frac{\left(\frac{3}{2} + 1 \right)^3}{3} - \frac{(-1 + 1)^3}{3} + \left(\frac{(4 - 4)^3}{3} - \frac{\left(\frac{3}{2} - 4 \right)^3}{3} \right)$

$= \frac{1}{3} \left(\frac{5}{2} \right)^3 - \frac{1}{3} \left(-\frac{5}{2} \right)^3$

$= \frac{1}{3} \left(\frac{125}{8} + \frac{125}{8} \right)$

$= \frac{1}{3} \times \frac{125}{4}$

$= \frac{125}{12} \text{ units}^2$

Question 17

a $\sin \frac{\pi c}{2} = -1$

$\frac{\pi c}{2} = \frac{3\pi}{2}$

So $c = 3$.

OR

$\cos \pi c = -1$

$\pi c = \pi, 3\pi, \ldots$

$\pi c = 3\pi$ (according to graph)

So $c = 3$.

b $A_1 = \int_0^{\frac{1}{3}} \cos \pi x - \sin \frac{\pi x}{2} dx$

$= \left[\frac{1}{\pi} \sin \pi x + \frac{2}{\pi} \cos \frac{\pi x}{2} \right]_0^{\frac{1}{3}}$

$= \frac{1}{\pi} \sin \frac{\pi}{3} + \frac{2}{\pi} \cos \frac{\pi}{6} - \left(0 + \frac{2}{\pi} \cos 0 \right)$

$= \frac{\sqrt{3}}{2\pi} + \frac{\cancel{2}}{\pi} \times \frac{\sqrt{3}}{\cancel{2}} - \frac{2}{\pi}$

$= \frac{\sqrt{3}}{2\pi} + \frac{\sqrt{3} - 2}{\pi}$

$A_2 = \int_{\frac{1}{3}}^{\frac{5}{3}} \sin \frac{\pi x}{2} - \cos \pi x \, dx$

$= \left[\frac{-2}{\pi} \cos \frac{\pi x}{2} - \frac{1}{\pi} \sin \pi x \right]_{\frac{1}{3}}^{\frac{5}{3}}$

$= \frac{-2}{\pi} \cos \frac{5\pi}{6} - \frac{1}{\pi} \sin \frac{5\pi}{3} - \left(\frac{-2}{\pi} \cos \frac{\pi}{6} - \frac{1}{\pi} \sin \frac{\pi}{3} \right)$

$= \frac{-2}{\pi} \left(\frac{-\sqrt{3}}{2} \right) - \frac{1}{\pi} \left(\frac{-\sqrt{3}}{2} \right) + \frac{2}{\pi} \times \frac{\sqrt{3}}{2} + \frac{1}{\pi} \times \frac{\sqrt{3}}{2}$

$= \frac{\sqrt{3}}{\pi} + \frac{\sqrt{3}}{\pi} + \frac{\sqrt{3}}{\pi}$

$= \frac{3\sqrt{3}}{\pi}$

$A_3 = \int_{\frac{5}{3}}^{3} \cos \pi x - \sin \frac{\pi x}{2} dx$

$= \left[\frac{-1}{\pi} \sin \pi x + \frac{2}{\pi} \cos \frac{\pi x}{2} \right]_{\frac{5}{3}}^{3}$

$= \frac{-1}{\pi} \sin 3\pi + \frac{2}{\pi} \cos \frac{3\pi}{2} - \left(\frac{-1}{\pi} \sin \frac{5\pi}{3} + \frac{2}{\pi} \cos \frac{5\pi}{6} \right)$

$= \frac{-1}{\pi} \left(\frac{-\sqrt{3}}{2} \right) + \frac{2}{\pi} \times \frac{\sqrt{3}}{2}$

$= \frac{\sqrt{3}}{\pi} + \frac{\sqrt{3}}{2\pi}$

Total area $= \frac{\sqrt{3}}{2\pi} + \frac{\sqrt{3}}{\pi} - \frac{2}{\pi} + \frac{3\sqrt{3}}{\pi} + \frac{\sqrt{3}}{\pi} + \frac{\sqrt{3}}{2\pi}$

$= \frac{6\sqrt{3}}{\pi} - \frac{2}{\pi}$

$= \frac{6\sqrt{3} - 2}{\pi} \text{ units}^2$

Question 18

a $y = \frac{1}{2} \log_e x$

$2y = \log_e x$

$x = e^{2y}$

b $x = 9, y = a$

$9 = e^{2a}$

$\ln 9 = 2a$

$a = \frac{1}{2} \ln 9$

$= \frac{1}{2} \ln 3^2$

$= \ln 3$

c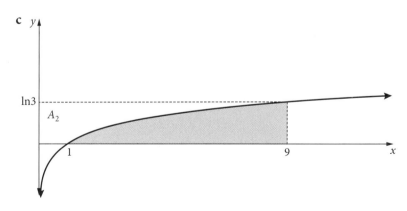

$A_1 \text{ (rectangle)} = 9 \times \ln 3 = 9 \ln 3$

$A_2 = \int_0^{\ln 3} e^{2y} \, dy$

$= 0.5(e^{2\ln 3} - e^0)$

$= 0.5(e^{\ln 9} - 1)$

$= 0.5(9 - 1)$

$= 4$

Shaded area $= (9 \ln 3 - 4) \text{ units}^2$

Question 19

$$\frac{dV}{dt} = \frac{2t}{1+t^2}$$

$V = \ln(1 + t^2) + c,$

$1 + t^2 > 0$ so absolute value is not required.

When $t = 0$, $V = 0$.

$0 = \ln(1 + 0^2) + c$

$\quad = 0 + c$

$c = 0$

So $V = \ln(1 + t^2)$.

When $t = 10$,

$V = \ln(1 + 10^2)$

$\quad = \ln 101$

The amount of water in the tank after 10 seconds is $\ln 101$ litres.

Question 20

a $v = 10 \times 5 - 5^2 = 25 \, \text{m/s}$

b $a = -4$

$v = -4t + c$

When $t = 10$, $v = 25$.

$25 = -4 \times 10 + c$

$\quad = -40 + c$

$c = 65$

$v = -4t + 65$

c Motorcyclist comes to rest when $v = 0$.

$0 = -4t + 65$

$4t = 65$

$t = 16.25$

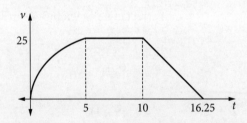

d $\int_0^5 10t - t^2 \, dt + 5 \times 25 + \int_{10}^{16.25} -4t + 65 \, dt$

$= \left[5t^2 - \frac{t^3}{3} \right]_0^5 + 125 + \left[-2t^2 + 65t \right]_{10}^{16.25}$

$= 125 - \frac{125}{3} - 0 + 125 - 2 \times (16.25)^2 + 65 \times 16.25 + 2 \times 10^2 - 65 \times 10$

$= 286.458\ldots$

$\approx 286 \, \text{m}$

HSC exam topic grid (2011–2020)

This table shows the coverage of this topic in past HSC exams by question number. The past exams can be downloaded from the NESA website (www.educationstandards.nsw.edu.au) by selecting 'Year 11 – Year 12', 'HSC exam papers'. NESA marking feedback and guidelines can also be found there.

Before 2020, 'Mathematics Advanced' was called 'Mathematics'. For these exams, select 'Year 11 – Year 12', 'Resources archive', 'HSC exam papers archive'.

	Integration	Trapezoidal rule	Areas under curves	Applications of integration
2011	2(e), 4(b), 4(d), 6(c)(ii)	5(c)*	6(c)	4(c)
2012	9, 11(g), 12(b)	12(d)	10, 13(b)	15(b)(iii)–(iv)[#]
2013	11(e)–(f)	15(a)(i)	13(b), 14(d)	14(a)[#], 16(a)
2014	4, 11(d)–(e), 13(a)	16(a)*	12(d)	11(f)
2015	11(g)–(h)	5	7, 9, 10, 16(a)	9[#], 14(a)[#], 15(c)
2016	9, 11(d), 12(d)	14(a)*	9, 13(e)	16(a)(iv)[#]
2017	11(b), 14(b)(i)	14(b)(ii)*	14(d)	9, **13(d)**, 15(c)[#], 16(b)[#]
2018	11(e)		7, 10, 15(b)–(c)(i)	
2019	9, 11(e), 13(c)	16(b)*	12(d), 16(c)	8[#], 14(a)[#], **14(b)(ii)**
2020 new course	4, 13, 17, **18**	20	7, 30	20

Questions in **bold** can be found in this chapter.
* Uses Simpson's rule, which is no longer in the course, but the trapezoidal rule can be used instead (with a similar answer)
[#] Motion

CHAPTER 5
SERIES, INVESTMENTS, LOANS AND ANNUITIES

9780170459228

SERIES, INVESTMENTS, LOANS AND ANNUITIES

Common content with Mathematics Standard 2 course

Arithmetic sequences and series

$T_n = a + (n - 1)d$

$T_n = T_{n-1} + d$

$S_n = \dfrac{n}{2}[2a + (n - 1)d]$

$S_n = \dfrac{n}{2}(a + l)$

Geometric sequences and series

$T_n = ar^{n-1}$

$T_n = rT_{n-1}$

$S_n = \dfrac{a(r^n - 1)}{r - 1}$

or $\quad S_n = \dfrac{a(1 - r^n)}{1 - r}$

$S_\infty = \dfrac{a}{1 - r}, \quad |r| < 1$

Investments

- Compound interest
- Effective interest rate

Reducing balance loans

- By tables and recurrence relations
- By geometric series

Annuities

- Future value
- Present value
- By tables and recurrence relations
- By geometric series

Glossary

annuity
An investment to which payments are made or received on a regular basis.

arithmetic sequence
A sequence of numbers such that the difference of any two consecutive terms is a constant (positive or negative).

arithmetic series
A sum of the terms of an arithmetic sequence.

common difference (*d*)
The same amount added (or subtracted) between consecutive terms in an arithmetic sequence.

common ratio (*r*)
The ratio between consecutive terms in a geometric sequence.

future value (FV) of an annuity
The total value of an annuity at the end of its term, including all interest earned.

future value interest factors
The values of an annuity of $1 at a specific time. A table of these factors can be used to calculate the future value of different amounts of money that are invested at a certain interest rate for a specified period of time.

general term, T_n
The *n*th term of a sequence for positive integer values of *n*. For example, if $T_n = n^2 + 1$, then $T_1 = 1^2 + 1 = 2$, $T_2 = 2^2 + 1 = 5$, $T_3 = 3^2 + 1 = 10$, … The first 3 terms of this sequence are 2, 5, 10.

A+ DIGITAL FLASHCARDS
Revise this topic's key terms and concepts by scanning the QR code or typing the URL into your browser.

https://get.ga/a-hsc-maths-advanced

geometric sequence
A sequence of numbers such that the ratio of any two consecutive terms is a constant (positive or negative).

geometric series
A sum of the terms of a geometric sequence.

present value (PV) of an annuity
The single sum of money (principal) that could be initially invested to produce the future value of an annuity over a given period of time.

present value (PV) interest factors
The present values of an annuity of $1 for a specific time. A table of these factors can be used to calculate the present value of different annuities that are invested at a certain interest rate for a specified period of time.

reducing balance loan
A compound interest loan that is repaid by making regular payments and the interest paid is calculated on the amount still owing (the reducing balance of the loan) after each payment has been made.

sequence
An ordered list of numbers whose terms follow a prescribed pattern. Mathematical sequences include arithmetic sequences and geometric sequences.

series
The sum of the terms of a particular sequence.

Topic summary

Modelling financial situations (MA-M1)

M1.2 Arithmetic sequences and series

The general term

In a sequence of terms that follow a pattern, the **general term** is represented by the notation T_n, where T_n is the nth term of the sequence.

Arithmetic sequences

In an **arithmetic sequence**, each term is a constant amount more (or less) than the previous term.

The constant amount is called the **common difference**, represented by d.

$$T_n = T_{n-1} + d \text{ so } d = T_2 - T_1 = T_3 - T_2 \text{ and generally, } d = T_n - T_{n-1}.$$

For example, 4, 10, 16, 22, … is an arithmetic sequence, where $T_1 = a = 4$ and $d = 6$ because

$$d = 10 - 4 = 16 - 10 = 6$$

The formula for the nth term in an arithmetic sequence with first term a and common difference d is

$$T_n = a + (n - 1)d.$$

So to find T_{12} for the example above,

$$T_{12} = 4 + (12 - 1)6$$
$$= 4 + 11 \times 6$$
$$= 70$$

Arithmetic series

The sum of the first n terms in an **arithmetic series** is

$$S_n = \frac{n}{2}[2a + (n - 1)d],$$

> **Hint**
> All T_n and S_n formulas appear on the HSC exam reference sheet and at the back of this book.

where a = first term, d = common difference.

Also,

$$S_n = \frac{n}{2}(a + l),$$

where a = first term, l = last term.

It can be useful to know that $T_n = S_n - S_{n-1}$.

Example 1

Find the sum of the first 12 terms of this arithmetic sequence: 4, 10, 16, 22, …

Solution

$a = 4, d = 10 - 4 = 6$

$$S_{12} = \frac{12}{2}\{2 \times 4 + (12 - 1) \times 6\}$$
$$= 6 \times \{8 + 11 \times 6\}$$
$$= 444$$

M1.3 Geometric sequences and series

Geometric sequences

In a **geometric sequence**, each term is a constant multiple of the previous term. The constant multiple is called the **common ratio**, represented by r.

For example, 5, –15, 45, –135, … is a geometric sequence, where $T_1 = a = 5$ and $r = \dfrac{-15}{5} = \dfrac{45}{-15} = -3$.

$$T_n = rT_{n-1} \quad \text{and} \quad r = \frac{T_n}{T_{n-1}}$$

The formula for each term in a geometric sequence with first term a and common ratio r is

$$T_n = ar^{n-1}.$$

Example 2

16, 8, 4, 2, … is a geometric sequence, where $a = 16$ and $r = \dfrac{8}{16} = \dfrac{4}{8} = \dfrac{2}{4} = \dfrac{1}{2}$.

Find the 10th term of this sequence.

Solution

$$T_n = ar^{n-1}$$
$$T_{10} = 16 \times \left(\frac{1}{2}\right)^{10-1}$$
$$= 16 \times \frac{1}{2^9}$$
$$= \frac{16}{512}$$
$$= \frac{1}{32}$$

Geometric series

The sum of a geometric series with n terms is given by

$$S_n = \frac{a(r^n - 1)}{r - 1} \quad \text{or} \quad S_n = \frac{a(1 - r^n)}{1 - r}.$$

> **Hint**
> Both of these formulas appear on the HSC exam reference sheet. The 2nd formula is more convenient to use if r is a fraction; that is, when $-1 < r < 1$.

Example 3

Find the sum of the first 10 terms of the geometric sequence in Example 2.

Solution

$$S_n = \frac{a(r^n - 1)}{r - 1}$$
$$S_{10} = \frac{16\left(\left(\frac{1}{2}\right)^{10} - 1\right)}{\frac{1}{2} - 1}$$
$$= \frac{16 \times \left(\frac{1}{1024} - 1\right)}{-\frac{1}{2}}$$
$$= -32 \times \left(-\frac{1023}{1024}\right)$$
$$= 31\frac{31}{32}$$

Limiting sum of an infinite geometric series

$$S_\infty = \frac{a}{1-r}, \text{ when } |r| < 1 \text{ (that is, when } -1 < r < 1, \text{ a fraction)}$$

Hint

The limiting sum does not require a value of n as it represents an infinite sum.

Example 4

Find the limiting sum of this infinite series:

$$3 + 1 + \frac{1}{3} + \frac{1}{9} + \ldots$$

Solution

This is a geometric series with $a = 3$, $r = \frac{1}{3}$.

It has a limiting sum because $|r| = \frac{1}{3} < 1$.

$$\begin{aligned}
S_\infty &= \frac{a}{1-r} \\
&= \frac{3}{1 - \frac{1}{3}} \\
&= \frac{3}{\frac{2}{3}} \\
&= 3 \times \frac{3}{2} \\
&= 4\frac{1}{2}
\end{aligned}$$

M1.1 Modelling investments and loans

Compound interest

Common content with Mathematics Standard 2 course

$$A = P(1 + r)^n,$$

where P = principal or **present value**

r = interest rate per period as a decimal

n = number of compounding periods

A = the final amount or **future value**

Note: $A = P + I$, so to calculate the amount of interest, $I = A - P$.

Effective annual rate of interest

This formula converts a compound interest rate that is not 'per year' to an equivalent interest rate per year.

$$\text{Effective annual rate of interest} = \left(1 + \frac{r}{n}\right)^n - 1,$$

where r = interest rate per period as a decimal

n = number of compounding periods per year

Hint

This formula allows comparisons of rates to be made when the effects of compounding over time are taken into account. It is not on the HSC reference sheet.

TOPIC SUMMARY

📎 Annuities

An **annuity** is a compound interest investment based on equal payments that are regularly deposited or received for a fixed period of time, such as superannuation.

The **present value of an annuity** is the single amount of money (equivalent to the principal) that *could* be invested to produce the same future value over the same time period.

The **future value of an annuity** is the total amount of the annuity at the end of the term and includes all contributions and interest earned.

📎 Future value of an annuity (FVA) – using a table of interest factors

This table gives the future value of an annuity (FVA) with a contribution of $1 at the end of each period at a given interest rate for a given time period. It can be used to calculate future values of annuities.

Future value interest factors for an annuity

Period	1%	2%	3%	4%	5%
1	1.0000	1.0000	1.0000	1.0000	1.0000
2	2.0100	2.0200	2.0300	2.0400	2.0500
3	3.0301	3.0604	3.0909	3.1216	3.1525
4	4.0604	4.1216	4.1836	4.2465	4.3101
5	5.1010	5.2040	5.3091	5.4163	5.5256
6	6.1520	6.3081	6.4684	6.6330	6.8019

Example 5

Use the table of future value interest factors above to find:

a the future value of an annuity of $7500 per year for 2 years at 4% p.a.

b the value of an investment of $2600 for 3 years at an interest rate of 4% p.a. compounded half-yearly.

Solution

a Interest rate p.a. = 4%
Total years = 2

In the table, this corresponds to 2.0400.

FV = $7500 × 2.04
   = $15 300

b Interest rate per half-year = 4% ÷ 2 = 2%
Total half-years = 3 × 2 = 6

In the table, this corresponds to 6.3081.

FV = $2600 × 6.3081
   = $16 401.06

Present value of an annuity (PVA) – using a table of interest factors

This table shows the present value of an annuity (PVA) with a contribution of $1 at the end of each period at a given interest rate over a given time period.

Present value interest factors for an annuity

Period	1%	2%	3%	4%	5%	6%
1	0.9901	0.9804	0.9709	0.9615	0.9524	0.9434
2	1.9704	1.9416	1.9135	1.8861	1.8594	1.8334
3	2.9410	2.8839	2.8286	2.7751	2.7232	2.6730
4	3.9020	3.8077	3.7171	3.6299	3.5460	3.4651
5	4.8534	4.7135	4.5797	4.4518	4.3295	4.2124

TOPIC SUMMARY

Example 6

a Boyd wants to invest $3600 per year in an annuity for 2 years at 4% p.a.

Calculate the present value of the annuity.

b Jemima wants to borrow $10 600 to be repaid over 5 years at 5% p.a.

Calculate her yearly repayment.

Solution

a Interest rate p.a. = 4%
Total years = 2

In the table, this corresponds to 1.8861.

$PV = \$3600 \times 1.8861$
$\quad = \$6789.96$

b Interest rate p.a. = 5%
Total years = 5

In the table, this corresponds to 4.3295.

If M represents the yearly repayment:

$M \times 4.3295 = \$10\,600$

$$M = \frac{\$10\,600}{4.3295}$$

$$= \$2448.32$$

M1.4 Financial applications of sequences and series

Annuities and geometric series

Example 7

An amount of $500 is invested at the beginning of each month.

If interest at 12% p.a. is paid monthly, how much is in the account at the end of 10 years?

> **Hint**
> Consider the structure of this problem carefully as you need to show all working out.

Solution

Consider these values: Monthly contribution = $500

r = 12% p.a. = 1% per month = 0.01

n = 10 years = 10 × 12 = 120 months

Let A_n be the amount in the account after n years.

The first contribution goes in at the start of the first month, so it earns interest for 120 months.

$A_1 = 500 \times 1.01^{120}$

The second contribution goes in at the start of the second month, so it earns interest for 119 months.

$A_2 = 500 \times 1.01^{119}$

The third contribution goes in at the start of the third month, so it earns interest for 118 months.

$A_3 = 500 \times 1.01^{118}$

$\vdots$

This pattern continues until the final contribution at the start of the 120th month.

$A_{120} = 500 \times 1.01^{1}$

$$FV = A_1 + A_2 + A_3 + \cdots + A_{120}$$
$$= 500 \times 1.01^{120} + 500 \times 1.01^{119} + 500 \times 1.01^{118} + \cdots + 500 \times 1.01^{2} + 500 \times 1.01$$
$$= 500(1.01^{120} + 1.01^{119} + 1.01^{118} + \cdots + 1.01^{2} + 1.01),$$

where the expression in brackets is a geometric series (reading right to left) with $a = 1.01$, $r = 1.01$, $n = 120$.

$$S_n = \frac{a(r^n - 1)}{r - 1}$$
$$S_{120} = \frac{1.01(1.01^{120} - 1)}{1.01 - 1}$$
$$= \frac{1.01(1.01^{120} - 1)}{0.01}$$
$$= \$232.3390\ldots$$

$$FV = 500 \times \$232.3390\ldots$$
$$= 116\,169.5382\ldots$$
$$\approx \$116\,169.54$$

After 10 years, there will be $116 169.54 in the investment account.

📎 Reducing balance loans

With a **reducing balance loan**, such as a home loan, the interest paid is calculated on the amount owing, so over time the interest should decrease.

Example 8

An amount of $450 000 is borrowed as a home loan over 25 years. It is repaid in monthly instalments and interest is charged at a rate of 3.6% p.a. Calculate the monthly repayment, M.

Solution

Consider these values: Loan amount = $450 000

$r = 3.6\%$ p.a. $= 3.6 \div 12 = 0.3\%$ per month $= 0.003$

$n = 25$ years $= 25 \times 12 = 300$ months

$M = ?$

Let A_n be the amount owing after n months.

1st month:

$A_1 = 450\,000 \times 1.003 - M$ adding 0.3% interest, subtracting the monthly repayment

2nd month:

$A_2 = A_1 \times 1.003 - M$
$= (450\,000 \times 1.003 - M) \times 1.003 - M$
$A_2 = 450\,000 \times 1.003^2 - 1.003M - M$

3rd month:

$A_3 = A_2 \times 1.003 - M$
$= (450\,000 \times 1.003^2 - 1.003M - M) \times 1.003 - M$
$= 450\,000 \times 1.003^3 - 1.003^2 M - 1.003M - M$
$A_3 = 450\,000 \times 1.003^3 - M(1.003^2 + 1.003 + 1)$

Continue this pattern:

$A_{300} = 450\,000 \times 1.003^{300} - M(1.003^{299} + 1.003^{298} + 1.003^{298} + \cdots + 1.003^2 + 1.003 + 1)$
$A_{300} = 0$ (loan fully repaid)
$0 = 450\,000 \times 1.003^{300} - M(1 + 1.003 + 1.003^2 + \cdots + 1.003^{238} + 1.003^{239})$,

where the expression inside the brackets is a geometric series with $a = 1$, $r = 1.003$, $n = 300$.

$$S_n = \frac{a(r^n - 1)}{r - 1}$$

$$S_{300} = \frac{1(1.003^{300} - 1)}{1.003 - 1}$$

$$= \frac{1.003^{300} - 1}{0.003}$$

$$= 485.4305\ldots$$

So $0 = 450\,000 \times 1.003^{300} - M(485.4305\ldots)$.

$$M(485.4305\ldots) = 450\,000 \times 1.003^{300}$$

$$M = \frac{450\,000 \times 1.003^{300}}{485.4305\ldots}$$

$$M = 2277.012\,206\ldots$$

$$\approx \$2277.02 \quad \text{rounding up}$$

The monthly repayment is $2277.02.

Practice set 1

Multiple-choice questions

Solutions start on page 123.

Question 1

The first 3 terms of an arithmetic sequence are 4, 7 and 10. What is the 12th term of this sequence?

A 37   **B** 40   **C** 246   **D** 264

Question 2

Which term of the sequence with nth term $T_n = 13 - 2n$ is equal to -21?

A 4th   **B** 5th   **C** 16th   **D** 17th

Question 3

The first 3 terms in a geometric sequence are 5, $\dfrac{5}{3}$ and $\dfrac{5}{9}$.

What is the 6th term in this sequence?

A $\dfrac{5}{729}$ **B** $\dfrac{5}{243}$ **C** $\dfrac{5}{27}$ **D** $\dfrac{5}{12}$

Question 4

What is the sum of the first 25 terms of the arithmetic series $8 + 10 + 12 + \ldots$?

A 56   **B** 58   **C** 800   **D** 825

Question 5

Which of the following is a term of the geometric series $1024 - 512 + 256 \ldots$?

A $-\dfrac{1}{2}$ **B** $-\dfrac{1}{4}$ **C** $\dfrac{1}{8}$ **D** 2

Question 6

Which of the following is a term of the geometric sequence $4p$, $-8p^2$, $16p^3$, $\ldots$?

A $512p^9$ **B** $-512p^9$ **C** $1024p^9$ **D** $-1024p^9$

Question 7

In a geometric sequence, $T_3 = 18$ and $T_6 = 486$, what is the first term, a, and the common ratio, r?

A $a = 2, r = 2$ **B** $a = 3, r = 2$ **C** $a = 3, r = 3$ **D** $a = 2, r = 3$

Question 8

Zoe's grandmother deposits an amount of money into an investment account for Zoe. It pays an annual rate of 3.6% p.a. compounded monthly.

How much interest is earned if Zoe's grandmother invests \$2500 for her over 2 years?

A \$15.02   **B** \$81.99   **C** \$186.35   **D** \$1725

Question 9 ●●●

The graph represents the future values of an investment invested at 3 different rates of compound interest for a number of years.

Which one of the following describes the compounding interest applied to each investment?

A X: 6% p.a. compounded annually
Y: 12% p.a. compounded annually
Z: 12% p.a. compounded monthly

B X: 12% p.a. compounded quarterly
Y: 12% p.a. compounded annually
Z: 6% p.a. compounded annually

C X: 6% p.a. compounded annually
Y: 12% p.a. compounded monthly
Z: 12% p.a. compounded quarterly

D X: 6% p.a. compounded monthly
Y: 12% p.a. compounded annually
Z: 12% p.a. compounded monthly

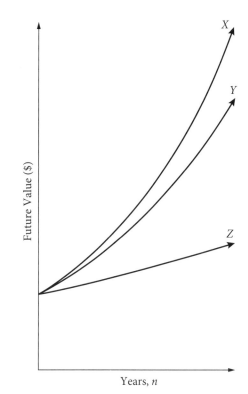

Question 10 ●●●

An arithmetic series has 28 terms. The first term is 8 and the last term is 170.

What is the sum of the series?

A 197   **B** 282   **C** 1947   **D** 2492

Question 11 ●●●

The fifth term of an arithmetic sequence is 27 and the eighth term is 12.

What is the common difference?

A −13   **B** −5   **C** 5   **D** 13

Question 12 ●●●

In an arithmetic series, $T_4 = 250$ and $S_6 = 1380$.

What is the first term, a, and common difference, d?

A $a = 130, d = 50$ **B** $a = 130, d = 40$ **C** $a = 140, d = 40$ **D** $a = 140, d = 20$

Question 13 ●●●

What is the sum of the first 8 terms of the geometric series $2 + 8 + 32 + \ldots$?

A $10\,922\frac{2}{3}$ **B** $43\,689\frac{2}{3}$ **C** $43\,690$ **D** $43\,690\frac{1}{3}$

Question 14 ●●●

For what values of x does the geometric series $1 + \dfrac{x}{5} + \dfrac{x^2}{25} + \dfrac{x^3}{125} + \ldots$ have a limiting sum?

A $-1 \leq x \leq 1$ **B** $-5 \leq x \leq 5$ **C** $-5 < x < 5$ **D** x can take any value

Question 15 ⬤⬤○

The sum of the first n terms of an arithmetic sequence is given by $S_n = \dfrac{n(5n + 3)}{2}$.

Which of the following represents the general term, T_n?

A $\quad T_n = \dfrac{n(5n + 3)}{2}$ 　　　B $\quad T_n = 9n - 5$ 　　　C $\quad T_n = 5n - 1$ 　　　D $\quad T_n = \dfrac{(n - 1)(5n - 2)}{2}$

Question 16 ⬤⬤○

Jude was comparing 4 loan rates as shown below.

Using the effective annual interest rate formula $\left(1 + \dfrac{r}{n}\right)^n - 1$, which loan offers Jude the best rate?
[Note: There are 365 days in a year.]

A　5.2% p.a. compounded daily 　　　　　　B　5.2% p.a. compounded monthly

C　4.9% p.a. compounded quarterly 　　　　D　4.9% p.a. compounded monthly

Question 17 ⬤⬤○

The table shows the future value of $1 for different interest rates and periods.

Future value interest factors

Period	Interest rate per period			
	2%	4%	6%	8%
1	1.00	1.00	1.00	1.00
2	2.02	2.04	2.06	2.08
3	3.06	3.12	3.18	3.25
4	4.12	4.25	4.37	4.51
5	5.20	5.42	5.64	5.87
6	6.31	6.63	6.98	7.34

What amount would need to be invested every quarter into an account earning 8% p.a. interest compounded quarterly, to be worth \$18 475 after 1 year?

A　\$4096 　　　　　B　\$4347 　　　　　C　\$4484 　　　　　D　\$9146

Question 18 ⬤⬤⬤

Alia invests a single amount of money into an account that earns interest compounding monthly.

The equivalent effective annual rate of interest is $\left(1 + \dfrac{r}{n}\right)^n - 1$, and for Alia's investment is 12.25%.

What is the interest rate p.a.?

A　8.41% 　　　　　B　10.45% 　　　　　C　11.61% 　　　　　D　12.25%

Question 19 ⬤⬤⬤

Find the limiting sum of the geometric series $2 + \dfrac{2}{\sqrt{2} + 1} + \dfrac{2}{(\sqrt{2} + 1)^2} + \ldots$

A　$\sqrt{2} + 1$ 　　　　B　$\sqrt{2} + 2$ 　　　　C　$\sqrt{2} + 3$ 　　　　D　$\sqrt{2} + 4$

Question 20 ⬤⬤⬤

The sum of the first 4 terms of an arithmetic sequence is 26. The sum of the first 12 terms is 222.

Calculate the sum of the first 20 terms.

A　610 　　　　　B　640 　　　　　C　1220 　　　　　D　1280

Practice set 2

Short-answer questions

Solutions start on page 126.

Question 1 (3 marks) ⬤◯◯

The nth term of a sequence is given by $T_n = 5^n$.

a Write down the first 3 terms in this sequence.   2 marks

b Which term of the sequence is $15\,625$?   1 mark

Question 2 (3 marks) ⬤◯◯

The nth term of a sequence is $2n^2$.

a Find the 3rd term of this sequence.   1 mark

b Is the number 300 a term of this sequence? Show working to justify your answer.   2 marks

Question 3 (2 marks) ⬤⬤◯

The first 4 terms of a sequence are $\ln 3$, $\ln 9$, $\ln 27$, $\ln 81$.

a What are the next 2 terms in this sequence?   1 mark

b Explain whether this series is arithmetic, geometric or neither.   1 mark
Show working to justify your answer.

Question 4 (4 marks) ⬤◯◯

For the series $3 - 6 + 12 - 24 + \ldots$, find:

a the common ratio, r.   1 mark

b the 8th term.   1 mark

c the sum of the first 12 terms.   2 marks

Question 5 (4 marks) ⬤◯◯

For the arithmetic sequence $\sqrt{3} + \sqrt{12} + \sqrt{27} + \ldots$

a find the common difference.   1 mark

b find an expression for the nth term.   1 mark

c calculate the sum of the first 10 terms.   2 marks

Question 6 (4 marks) ⬤◯◯

An arithmetic sequence is given by $2x + 1$, $3x + 4$, $4x + 7$, $5x + 10$, $\ldots$

a Find the common difference, d.   2 marks

b Find an expression for T_{12}.   1 mark

c Write an expression for the sum of the first 25 terms of the series.   1 mark

Question 7 (2 marks) ▣

An infinite geometric series has a first term of 2 and a limiting sum of $1\frac{1}{5}$.

What is its common ratio?   2 marks

Question 8 (2 marks) ▣

A tennis ball is dropped from a height of 5 metres onto a concrete floor. After its first bounce, it rises to a height of 2 metres and continues to bounce to $\frac{2}{3}$ of its previous height, until it stops.

What is the total distance through which the tennis ball travels?   2 marks

Question 9 (2 marks) ▣

Can there exist an infinite geometric series with a limiting sum, $S_\infty = \frac{7}{9}$, and a first term of 3?   2 marks
Show all working to justify your answer.

Question 10 (4 marks) ▣

Owen is training for a charity fun run and runs every week for 14 weeks. He runs 2 km in the first week and each week after that he runs 800 metres more than the previous week, until he can run 6 km in a single week. He then continues running 6 km each week.

a   How far does Owen run in the 4th week?   1 mark

b   In which week does he run 6 km?   1 mark

c   How far will Owen run over the 14 weeks of training?   2 marks

Question 11 (2 marks) ▣

The table below shows the present value of an annuity of $1.

Present value interest factors for an annuity

Period	4%	5%	6%	7%	8%	9%	10%
1	0.9615	0.9524	0.9434	0.9346	0.9259	0.9174	0.9091
2	1.8861	1.8594	1.8334	1.8080	1.7833	1.7591	1.7355
3	2.7751	2.7232	2.6730	2.6243	2.5771	2.5313	2.4869
4	3.6299	3.5460	3.4651	3.3872	3.3121	3.2397	3.1699
5	4.4518	4.3295	4.2124	4.1002	3.9927	3.8897	3.7908
6	5.2421	5.0757	4.9173	4.7665	4.6229	4.4859	4.3553
7	6.0021	5.7864	5.5824	5.3893	5.2064	5.0330	4.8684
8	6.7327	6.4632	6.2098	5.9713	5.7466	5.5348	5.3349
9	7.4353	7.1078	6.8017	6.5152	6.2469	5.9952	5.7590
10	8.1109	7.7217	7.3601	7.0236	6.7101	6.4177	6.1446
11	8.7605	8.3064	7.8869	7.4987	7.1390	6.8052	6.4951
12	9.3851	8.8633	8.3838	7.9427	7.5361	7.1607	6.8137

Use the table above to answer the following questions.

a   Vakul takes out a personal loan of $12 000, which is to be repaid over 5 years at an interest rate of 7% p.a. Calculate Vakul's yearly repayments.   1 mark

b   Kemala plans to invest $2000 twice a year for 4 years in an annuity. Her investment will earn interest of 10% p.a. What is the present value of Kemala's annuity?   1 mark

Question 12 (5 marks)

Bernadette has a large water storage tank on her farm. When she fully turns on the tap at the bottom of the tank, 84 L pours out in the first minute, 78 L in the second minute and 72 L in the third minute. Assume this pattern continues and no water is added, until the tank is completely empty.

a How many litres of water pour out in the 9th minute?   1 mark

b How much water, in litres, has poured out after 11 minutes?   2 marks

c Determine the capacity of this water storage tank.   2 marks

Question 13 (5 marks)

An alpine expedition starts from a base camp situated at an altitude of 1200 m. After one day's climbing on the mountain, the trekkers increase their altitude by 1000 m. Each subsequent day's climb increases their altitude by $\frac{2}{3}$ of the increase of the previous day.

a Calculate, to the nearest metre, the altitude reached after 8 days of climbing.   3 marks

b Can the expedition reach a summit of altitude 4300 m? Show all working to justify   2 marks
your answer.

Question 14 (3 marks) ©NESA 2020 HSC EXAM, QUESTION 12

Calculate the sum of the arithmetic series $4 + 10 + 16 + \cdots + 1354$.   3 marks

Question 15 (5 marks)

The diagram shows the seating area for an outdoor theatre.

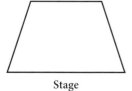

Stage

The first row has 51 seats. Each row of seats has 2 fewer seats than the row directly in front of it.

a Write an expression for T_n to represent the number of seats in the nth row.   1 mark

b Hence, find the maximum number of rows of seats possible in the outdoor theatre.   2 marks

c Given that the outdoor theatre holds 352 people in total, how many rows of seats are there? 2 marks

Question 16 (5 marks)

A loan of \$350 000 is to be repaid in equal monthly instalments of \$M over a period of 20 years. Interest is charged each month at a rate of 4.5% p.a. It is calculated on the balance owing at the beginning of the month and added to the balance. A_n represents the amount owing after n months.

a Show that $A_1 = 350\,000 \times 1.003\,75 - M$.   1 mark

b Show that $A_3 = 350\,000 \times 1.003\,75^3 - M(1 + 1.003\,75 + 1.003\,75^2)$.   2 marks

c Calculate the value of M, correct to the nearest cent.   2 marks

Question 17 (5 marks)

A furniture store charges 7.5% p.a. interest, compounded monthly, for loans. Repayments do not have to be made for the first 3 months. Dominic buys \$12 500 worth of furniture and pays it off over 48 months.

a How much does Dominic owe after 3 months?   1 mark

b Construct the recurrence relation for this loan, where A_0 is the amount borrowed   2 marks
and A_n represents the amount owing after n months and, hence, find the amount of
each monthly repayment, M.

c How much does Dominic repay altogether?   1 mark

d How much interest does Dominic pay altogether?   1 mark

PRACTICE SET 2

Question 18 (7 marks) ●●●

A wildlife conservancy finds that a koala population is decreasing at a rate of 12% each year. The koala population was 50 000 in November 2021.

a Write an equation to represent the number of koalas remaining after n years. 2 marks

b Hence, calculate the number of koalas predicted to remain after 6 years. 2 marks

c In what year and month is the koala population expected to be 32 000? 3 marks

Question 19 (7 marks) ©NESA 2020 HSC EXAM, QUESTION 26 ●●●

Tina inherits $60 000 and invests it in an account earning interest at a rate of 0.5% per month. Each month, immediately after the interest has been paid, Tina withdraws $800.

The amount in the account immediately after the nth withdrawal can be determined using the recurrence relation

$$A_n = A_{n-1}(1.005) - 800,$$

where $n = 1, 2, 3, \ldots$ and $A_0 = 60 000$.

a Use the recurrence relation to find the amount of money in the account immediately after the third withdrawal. 2 marks

b Calculate the amount of interest earned in the first three months. 2 marks

c Calculate the amount of money in the account immediately after the 94th withdrawal. 3 marks

Question 20 (3 marks) ●●●

Evaluate the sum of the first 15 terms of the series $4 + 9 + 16 + 27 + 46 + \ldots$ 3 marks

Practice set 1

Worked solutions

1 A

4, 7, 10, ...

$a = 4, d = 7 - 4 = 3$

$T_n = a + (n - 1)d$
$T_{12} = 4 + 11 \times 3$
$\quad\ = 4 + 33$
$\quad\ = 37$

2 D

$-21 = 13 - 2n$
$-34 = -2n$
$\quad\ n = 17$

3 B

$5, \dfrac{5}{3}, \dfrac{5}{9}$

$r = \dfrac{5}{3} \div 5 = \dfrac{\cancel{5}}{3} \times \dfrac{1}{\cancel{5}} = \dfrac{1}{3}$

$T_n = ar^{n-1}$

$T_6 = 5 \times \left(\dfrac{1}{3}\right)^5$

$\quad\ = 5 \times \dfrac{1}{3^5}$

$\quad\ = \dfrac{5}{243}$

4 C

$a = 8, d = 2, S_{25} = ?$

$S_n = \dfrac{n}{2}[2a + (n - 1)d]$

$S_{25} = \dfrac{25}{2}[2 \times 8 + 24 \times 2]$

$\quad\ = 12.5 \times 64$

$\quad\ = 800$

5 A

$a = 1024, r = -\dfrac{1}{2}$

$T_3 = 256 \times \left(-\dfrac{1}{2}\right) = -128$

$T_4 = -128 \times \left(-\dfrac{1}{2}\right) = 64$

$T_5 = 64 \times \left(-\dfrac{1}{2}\right) = -32$

Then $16, -8, 4, -2, 1, -\dfrac{1}{2}$.

6 C

$a = 4p$

$r = \dfrac{-8p^2}{4p} = -2p$

$T_n = 4p \times (-2p)^{n-1}$

For p^9 in the term, we want $n = 9$:

$T_9 = 4p \times (-2p)^8$
$\quad\ = 4p \times 256p^8$
$\quad\ = 1024p^9$

7 D

$T_3 = ar^2 = 18$
$T_6 = ar^5 = 486$

$\dfrac{ar^5}{ar^2} = \dfrac{486}{18}$

$\quad r^3 = 27$

$\quad\ r = 3$

By substituting $r = 3$ into $ar^2 = 18$, $a = 2$.

8 C

$2500 \times 1.003^{24} = 2686.348\ldots$
$2686.348\ldots - 2500 = \$186.35$

9 B

X: 12% compounded quarterly (compounding more frequently gives a little higher FV than compounding annually)

Y: 12% compounded annually

Z: 6% compounded annually

10 D

$n = 28, a = 8, l = 170$

$S_n = \dfrac{n}{2}(a + l)$

$S_{28} = \dfrac{28}{2}(8 + 170)$

$\quad\ = 2492$

11 B

$T_5 = 27 = a + 4d$ $\qquad$ [1]
$T_8 = 12 = a + 7d$ $\qquad$ [2]

[1] − [2]:

$27 - 12 = 4d - 7d$
$\qquad 15 = -3d$
$\qquad\quad d = -5$

12 B

$T_4 = 250 = a + 3d$ $\qquad$ [1]
$S_6 = 1380 = 3(2a + 5d)$
$460 = 2a + 5d$ $\qquad$ [2]

Rearrange [1]:

$a = 250 - 3d$ $\qquad$ [3]

Substitute [3] into [2]:

$460 = 2(250 - 3d) + 5d$
$460 = 500 - 6d + 5d$
$-40 = -d$
$\quad d = 40$

Substitute $d = 40$ into [1]:

$250 = a + 3 \times 40$
$250 = a + 120$
$\quad a = 130$

13 C

$a = 2, r = 4$

$S_8 = \dfrac{2(4^8 - 1)}{4 - 1}$
$\quad = 43\,690$

14 C

$r = \dfrac{x}{5} \qquad |r| < 1$

$-1 < \dfrac{x}{5} < 1$

$-5 < x < 5$

15 C

$S_1 = \dfrac{5 + 3}{2} = 4$

$S_2 = \dfrac{2(2 \times 5 + 3)}{2} = 13$

$S_3 = \dfrac{3(3 \times 5 + 3)}{2} = \dfrac{3 \times 18}{2} = 27$

$S_1 = T_1 = 4$

$S_1 + T_2 = 13$
$\qquad T_2 = 13 - 4$
$\qquad T_2 = 9$

$T_3 = 27 - 13$
$\quad\ = 14$

$a = 4, d = 5 \qquad 4, 9, 14, \dots$

$T_n = a + (n - 1)d$
$T_n = 4 + 5n - 5$
$T_n = 5n - 1$

16 A is highest.

A: $\left(1 + \dfrac{0.052}{365}\right)^{365} - 1 = 0.053\,37\dots$

B: $\left(1 + \dfrac{0.052}{12}\right)^{12} - 1 = 0.053\,25\dots$

C: $\left(1 + \dfrac{0.049}{4}\right)^{4} - 1 = 0.0499\dots$

D: $\left(1 + \dfrac{0.049}{12}\right)^{12} - 1 = 0.050\,11\dots$

17 C

8% ÷ 4 = 2% per quarter

2% for 4 quarters, read off the table: 4.12

$\quad FV = 4.12 \times PV$
$18\,475 = 4.12 \times PV$
$\quad PV = \dfrac{18\,475}{4.12}$
$\qquad = \$4484$

18 C

$\left(1 + \dfrac{r}{12}\right)^{12} - 1 = 0.1225$

$\left(1 + \dfrac{r}{12}\right)^{12} = 1.1225$

$1 + \dfrac{r}{12} = 1.009\,676\,378$

$\dfrac{r}{12} = 0.009\,676\,378$

$r = 0.116\,116\dots$

$\approx 11.61\%$

19 B

$$a = 2$$

$$r = \frac{2}{\sqrt{2}+1} \div 2$$

$$= \frac{1}{\sqrt{2}+1}$$

$$S_\infty = \frac{a}{1-r}$$

$$= \frac{2}{1-\left(\dfrac{1}{\sqrt{2}+1}\right)}$$

$$= \frac{2}{\left(\dfrac{\sqrt{2}+1-1}{\sqrt{2}+1}\right)}$$

$$= 2 \times \frac{\sqrt{2}+1}{\sqrt{2}}$$

$$= \frac{2\sqrt{2}+2}{\sqrt{2}} \times \frac{\sqrt{2}}{\sqrt{2}}$$

$$= \frac{4+2\sqrt{2}}{2}$$

$$= 2+\sqrt{2}$$

20 A

$$S_4 = 26 = 2(2a + 3d)$$
$$13 = 2a + 3d \qquad [1]$$

$$S_{12} = 222 = 6(2a + 11d)$$
$$37 = 2a + 11d \qquad [2]$$

$$[2] - [1]:$$

$$37 - 13 = 8d$$
$$24 = 8d$$
$$d = 3, \text{ so } a = 2$$

$$S_n = \frac{n}{2}[2a + (n-1)d]$$

$$S_{20} = \frac{20}{2}(2 \times 2 + 19 \times 3)$$

$$= 610$$

WORKED SOLUTIONS

Practice set 2

Worked solutions

Question 1

a $T_1 = 5$
$T_2 = 5^2 = 25$
$T_3 = 5^3 = 125$

b $5^n = 15\,625$
$5^n = 5^6$ (or by using logarithms)
So $n = 6$.

Question 2

a $T_3 = 2 \times 3^2$
$\quad = 18$

b $300 = 2n^2$
$150 = n^2$
$n = \pm\sqrt{150} \qquad n > 0$
$\quad = 12.247\ldots$

n is not a positive integer so 300 is not a term in this sequence.

Question 3

a $\ln 3, \ln 9, \ln 27, \ln 81$
$\ln 3, \ln 3^2, \ln 3^3, \ln 3^4$
$\ln 3, 2\ln 3, 3\ln 3, 4\ln 3$

So the next 2 terms are $5\ln 3$ and $6\ln 3$, where
$5\ln 3 = \ln 3^5 = \ln 243$ and $6\ln 3 = \ln 3^6 = \ln 729$

b It is an arithmetic series as
$T_2 - T_1 = 2\ln 3 - \ln 3 = \ln 3$
and
$T_3 - T_2 = 3\ln 3 - 2\ln 3 = \ln 3$.

So the common difference, d, is $\ln 3$.

Question 4

a $a = 3, r = \dfrac{-6}{3} = \dfrac{12}{-6} = -2$

b $T_n = ar^{n-1}$
$T_8 = 3 \times (-2)^7$
$\quad = -384$

c $S_n = \dfrac{a(r^n - 1)}{r - 1}$

$S_{12} = \dfrac{3((-2)^{12} - 1)}{-2 - 1}$

$\quad = -4095$

Question 5

a $\sqrt{3} + \sqrt{12} + \sqrt{27} + \ldots = \sqrt{3} + 2\sqrt{3} + 3\sqrt{3} + \ldots$
$2\sqrt{3} - \sqrt{3} = 3\sqrt{3} - 2\sqrt{3} = \sqrt{3}$
So $d = \sqrt{3}$.

b $T_n = \sqrt{3} + (n - 1)\sqrt{3}$
$\quad = \sqrt{3} + \sqrt{3}n - \sqrt{3}$
$T_n = \sqrt{3}n$

c $S_n = \dfrac{n}{2}[2a + (n - 1)d]$

$S_{10} = \dfrac{10}{2}[2\sqrt{3} + 9 \times \sqrt{3}]$

$\quad = 5 \times 11\sqrt{3}$

$\quad = 55\sqrt{3}$

Question 6

a $3x + 4 - (2x + 1) = x + 3$
$4x + 7 - (3x + 4) = 4x + 7 - 3x - 4 = x + 3$
So $\qquad d = x + 3$

b $T_n = (2x + 1) + (n - 1)(x + 3)$
$T_{12} = (2x + 1) + 11(x + 3)$
$\quad = 2x + 1 + 11x + 33$
$T_{12} = 13x + 34$

c $S_n = \dfrac{n}{2}[2a + (n - 1)d]$

$S_{25} = \dfrac{25}{2}[2(2x + 1) + 24(x + 3)]$

$\quad = \dfrac{25}{2}(4x + 2 + 24x + 72)$

$\quad = 350x + 925$

Question 7

$$S_\infty = \frac{a}{1-r}, a = 2, S_\infty = 1\frac{1}{5}, r = ?$$

$$1\frac{1}{5} = \frac{2}{1-r}$$

$$\frac{6}{5} = \frac{2}{1-r}$$

$$6(1-r) = 10$$

$$6 - 6r = 10$$

$$-6r = 4$$

$$r = -\frac{2}{3}$$

Question 8

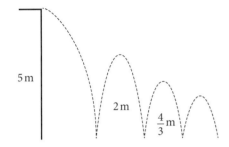

After the ball drops 5 metres, it completes each bounce from the ground to its new height and back to the ground an infinite number of times until it stops.

$$S_\infty = \frac{2}{1 - \frac{2}{3}} \times 2$$

$$= \frac{2}{\frac{1}{3}} \times 2$$

$$= 6 \times 2$$

$$= 12$$

The total distance travelled by the ball is $5 + 12 = 17$ m.

Question 9

$$S_\infty = \frac{a}{1-r}$$

$$\frac{7}{9} = \frac{3}{1-r}$$

$$7 - 7r = 27$$

$$-7r = 20$$

$$r = -\frac{20}{7} = -2\frac{6}{7}$$

No, $|r| < 1$ for a limiting sum.

Question 10

$a = 2\,\text{km}$

$d = 0.8\,\text{km}$

a $T_4 = 2 + 3 \times 0.8$

   $= 4.4\,\text{km}$

b $6 = 2 + (n-1)0.8$

   $= 2 + 0.8n - 0.8$

   $= 1.2 + 0.8n$

   $4.8 = 0.8n$

   $n = 6$

Owen runs 6 km in the 6th week.

c $S_6 = 3[2 \times 2 + (n-1)0.8]$

   $= 3(4 + 5 \times 0.8)$

   $= 24$

$14 - 6 = 8$ weeks

$8 \times 6 = 48$

$24 + 48 = 72\,\text{km}$

Owen will run 72 km over the 14 weeks of training.

Question 11

a 5 years, 7%, read off table = 4.1002

$12\,000 = 4.1002 \times M$, where M = yearly repayment

$$M = \frac{12\,000}{4.1002} = \$2926.69$$

Vakul's yearly repayments are $2926.69.

b 10% p.a. ÷ 2 = 5% per half-year

4 years, $4 \times 2 = 8$ payments

Read off table = 6.4632

$PV = 2000 \times 6.4632$

   $= \$12\,926.40$

The present value of Kemala's annuity is $12 926.40.

Question 12

$84 + 78 + 72 + \ldots$

$a = 84, d = -6$

a $T_n = a + (n - 1)d$
$T_9 = 84 + 8 \times (-6)$
$\quad = 36\,\text{L}$

b $S_n = \dfrac{n}{2}[2a + (n - 1)d]$

$S_{11} = \dfrac{11}{2}[2 \times 84 + 10 \times (-6)]$

$\quad = 594\,\text{L}$

c $T_n = 84 + (n - 1)(-6)$
$\quad = 84 - 6n + 6$
$\quad = 90 - 6n$

When $T_n = 0$, the tank is empty.

$90 - 6n = 0$
$\qquad n = 15$

When $n = 15$, the tank is empty.

$S_n = \dfrac{n}{2}[2a + (n - 1)d]$

$S_{15} = \dfrac{15}{2}[2 \times 84 + 14 \times (-6)]$

$\quad = 630\,\text{L}$

The tank has a capacity of 630 L.

Question 13

a Base camp altitude = 1200 m

$a = 1000, r = \dfrac{2}{3}$

$S_n = \dfrac{a(r^n - 1)}{r - 1}$

$S_8 = \dfrac{1000\left(\left(\frac{2}{3}\right)^8 - 1\right)}{\frac{2}{3} - 1}$

$\quad = 2882.944\,673$

Altitude = 1200 + 2882.944 673
$\qquad\quad = 4082.944\,673$

Altitude of 4083 m after 8 days.

b $1200 + \dfrac{1000}{1 - \frac{2}{3}} = 4200\,\text{m} \ < 4300\,\text{m}$

No, the summit is beyond reach.

Question 14

$4 + 10 + 16 + \cdots + 1354$

$a = 4, d = 6$

$T_n = a + (n - 1)d$

$1354 = 4 + 6(n - 1)$
$\quad\ \ = 6n - 2$
$1356 = 6n$
$\quad\ n = 226$

$S_n = \dfrac{n}{2}(a + l)$

$S_{226} = \dfrac{226}{2}(4 + 1354)$

$\quad = 153\,454$

Question 15

a $a = 51, d = -2$

$T_n = a + (n - 1)d$
$\quad = 51 + (n - 1)(-2)$
$\quad = 51 - 2n + 2$
$T_n = 53 - 2n$

b $53 - 2n > 0$
$\quad\ 53 > 2n$
$\quad \dfrac{53}{2} > n$
$\qquad n < 26.5$
$\qquad n = 26$

The maximum number of rows of seats is 26.

c $352 = \dfrac{n}{2}[2 \times 51 + (n - 1) \times (-2)]$

$352 \times 2 = n(102 - 2n + 2)$

$704 = n(104 - 2n)$

$704 = 104n - 2n^2$

$2n^2 - 104n + 704 = 0$

$n^2 - 52n + 352 = 0$

$(n - 44)(n - 8) = 0$

$n = 8, 44$ but $n < 26$

So $n = 8$.

Question 16

a $350\,000$ loan

M monthly instalments

$4.5\% \div 12 = 0.003\,75$

$20 \times 12 = 240$ instalments

1st month: $A_1 = 350\,000 \times (1 + 0.003\,75) - M$

$\qquad = 350\,000 \times 1.003\,75 - M$

b 2nd month: $A_2 = A_1 \times 1.003\,75 - M$

$\qquad\qquad = (350\,000 \times 1.003\,75 - M) \times 1.003\,75 - M$

$\qquad A_2 = 350\,000 \times 1.003\,75^2 - 1.003\,75M - M$

3rd month: $A_3 = A_2 \times 1.003\,75 - M$

$\qquad A_3 = (350\,000 \times 1.003\,75^2 - 1.003\,75M - M) \times 1.003\,75 - M$

$\qquad\qquad = 350\,000 \times 1.003\,75^3 - 1.003\,75^2 M - 1.003\,75M - M$

$\qquad A_3 = 350\,000 \times 1.003\,75^3 - M(1.003\,75^2 + 1.003\,75 + 1)$

c 20 years = 240 months

Continuing the pattern:

$A_{240} = 350\,000 \times 1.003\,75^{240} - M(1.003\,75^{239} + 1.003\,75^{238} + \cdots + 1.003\,75^2 + 1.003\,75 + 1)$

$A_{240} = 0$ (loan fully repaid)

$0 = 350\,000 \times 1.003\,75^{240} - M(1 + 1.003\,75 + 1.003\,75^2 + \cdots + 1.003\,75^{238} + 1.003\,75^{239})$

The expression in brackets is a geometric series with $a = 1$, $r = 1.003\,75$, $n = 240$.

$$S_n = \frac{a(r^n - 1)}{r - 1}$$

$$S_{240} = \frac{1.003\,75^{240} - 1}{1.003\,75 - 1}$$

$$= \frac{1.003\,75^{240} - 1}{0.003\,75}$$

$$= 388.124\,362\,9\ldots$$

So $0 = 350\,000 \times 1.003\,75^{240} - M(388.124\,362\,9\ldots)$.

$M(388.124\,362\,9\ldots) = 350\,000 \times 1.003\,75^{240}$

$$M = \frac{350\,000 \times 1.003\,75^{240}}{388.124\,362\,9\ldots}$$

$$= 2214.272\,817\ldots$$

$$= \$2214.28 \quad \text{(rounding up)}$$

Question 17

There are no repayments for the first 3 months.

7.5% p.a. $\div 12 = 0.625\%$ per month $= 0.006\,25$

Loan $= \$12\,500$ for 48 months

Interest is paid for $48 - 3 = 45$ months only.

a Use compound interest formula for 3 months:

$A = 12\,500(1 + 0.006\,25)^3$

$\quad = 12\,500 \times 1.006\,25^3$

$\quad \approx \$12\,735.84$

b $A_3 = 12\,500 \times 1.006\,25^3$

$A_4 = A_3 \times 1.006\,25 - M$
$\quad = 12\,500 \times 1.006\,25^4 - M$

$A_5 = A_4 \times 1.006\,25 - M$
$\quad = (12\,500 \times 1.006\,25^4 - M) \times 1.006\,25 - M$
$\quad = 12\,500 \times 1.006\,25^5 - 1.006\,25M - M$

$A_6 = 12\,500 \times 1.006\,25^6 - 1.006\,25^2 M - 1.006\,25M - M$

$\vdots$

$A_{48} = 12\,500 \times 1.006\,25^{48} - M(1.006\,25^{44} + 1.006\,25^{43} + 1.006\,25^{42} + \cdots + 1.006\,25 + 1)$
$A_{48} = 0$ (loan fully repaid)

$$0 = 12\,500 \times 1.006\,25^{48} - M\left(\frac{1.006\,25^{45} - 1}{0.006\,25}\right)$$

$M(51.780\,117\ldots) = 12\,500 \times 1.006\,25^{48}$
$$M = \frac{12\,500 \times 1.006\,25^{48}}{51.780\,117\ldots}$$
$$= 325.559\,11\ldots$$
$$\approx \$325.56$$

c $325.56 \times 45 = \$14\,647.50$

d $14\,647.50 - 12\,500 = \$2147.50$

Question 18

a Koala population is in decline: $100 - 12 = 88\%$

So $r = 0.88$.

So $T_n = 50\,000 \times 0.88^n$.

b $T_6 = 50\,000 \times 0.88^6$
$\quad \approx 23\,220$

c $32\,000 = 50\,000 \times 0.88^n$

$\dfrac{32\,000}{50\,000} = 0.88^n$

$0.64 = 0.88^n$

$\ln 0.64 = \ln 0.88^n$

$\ln 0.64 = n \ln 0.88$

$n = \dfrac{\ln 0.64}{\ln 0.88}$
$\quad = 3.49$ years
$\quad \approx 3$ years 6 months

To reach a population of $32\,000$:

November 2021 + 3 years 6 months = May 2025

WORKED SOLUTIONS

Question 19

a $60\,000$ invested

0.5% per month = 0.005

$800 withdrawn per month

$A_1 = A_0 \times 1.005 - 800$
$A_1 = 60\,000 \times 1.005 - 800$
$\quad = 59\,500$

$A_2 = A_1 \times 1.005 - 800$
$\quad = 59\,500 \times 1.005 - 800$
$\quad = 58\,997.50$

$A_3 = A_2 \times 1.005 - 800$
$\quad = 58\,997.50 \times 1.005 - 800$
$\quad = 58\,492.4875$
$\quad \approx \$58\,492.49$

b 1st month: $\quad I = 60\,000 \times 0.005 = \$300$

2nd month: $\quad I = 59\,500 \times 0.005 = \$297.50$

3rd month: $\quad I = 58\,997.50 \times 0.005 = 294.9875 = \$294.99$

Total interest = $300 + $297.50 + $294.99 = $892.49

c $A_1 = 60\,000 \times 1.005 - 800$

$A_2 = (60\,000 \times 1.005 - 800) \times 1.005 - 800$
$\quad = 60\,000 \times 1.005^2 - 800 \times 1.005 - 800$
$\quad = 60\,000 \times 1.005^2 - 800(1.005 + 1)$

$A_3 = [60\,000 \times 1.005^2 - 800(1.005 + 1)] \times 1.005 - 800$
$\quad = 60\,000 \times 1.005^3 - 800 \times 1.005^2 - 800 \times 1.005 - 800$
$\quad = 60\,000 \times 1.005^3 - 800(1.005^2 + 1.005 + 1)$

Continuing the pattern:

$A_{94} = 60\,000 \times 1.005^{94} - 800(1.005^{93} + 1.005^{92} + \cdots + 1.005^2 + 1.005 + 1)$

$\quad = 60\,000 \times 1.005^{94} - 800\left(\dfrac{1.005^{94} - 1}{1.005 - 1}\right)$

$\quad = 60\,000 \times 1.005^{94} - 800\left(\dfrac{1.005^{94} - 1}{0.005}\right)$

$\quad = 60\,000 \times 1.005^{94} - 160\,000(1.005^{94} - 1)$
$\quad = 187.845\,997\,8\ldots$

Amount of money in account straight after 94th withdrawal:

$A_{94} = \$187.85$

Question 20

$4 + 9 + 16 + 27 + 46 + \ldots$

n	T_n	$T_n = 2^n$ (geometric)	$T_n = 3n - 1$ (arithmetic)	
$n = 1$	4	$2 = 2^1$	$2 = 3 \times 1 - 1$	$2 + 2 = 4$
$n = 2$	9	$4 = 2^2$	$5 = 3 \times 2 - 1$	$4 + 5 = 9$
$n = 3$	16	$8 = 2^3$	$8 = 3 \times 3 - 1$	$8 + 8 = 16$
$n = 4$	27	$16 = 2^4$	$11 = 3 \times 4 - 1$	$16 + 11 = 27$
$n = 5$	46	$32 = 2^5$	$14 = 3 \times 5 - 1$	$32 + 14 = 46$

$T_n = 2^n + 3n - 1$, which is a sum of a geometric and an arithmetic sequence.

Sum of geometric sequence for the first 15 terms:

$$S_n = \frac{a(r^n - 1)}{r - 1}, \text{ where } a = 2, r = 2$$

$$S_{15} = \frac{2(2^{15} - 1)}{2 - 1}$$

$$= 65\,534$$

Sum of arithmetic sequence for the first 15 terms:

$$S_n = \frac{n}{2}[2a + (n - 1)d], \text{ where } a = 2, d = 3$$

$$S_{15} = \frac{15}{2}[2 \times 2 + 14 \times 3]$$

$$= 345$$

The total sum (geometric and arithmetic)

$= 65\,534 + 345$

$= 65\,879$

HSC exam topic grid (2011–2020)

This table shows the coverage of this topic in past HSC exams by question number. The past exams can be downloaded from the NESA website (www.educationstandards.nsw.edu.au) by selecting 'Year 11 – Year 12', 'HSC exam papers'. NESA marking feedback and guidelines can also be found there.

Before 2020, 'Mathematics Advanced' was called 'Mathematics'. For these exams, select 'Year 11 – Year 12', 'Resources archive', 'HSC exam papers archive'.

	Arithmetic sequences and series	Geometric sequences and series	Loans and annuities by series	Investments, loans and annuities (introduced in 2020)	
2011	3(a)	5(a)	8(c)	22, 23(c)	General Maths exams
2012	12(c)	15(a)	15(c)	9, 24	
2013		12(c)	13(d)	9, 26(e)	
2014	12(a), 14(d)	8	16(b)	21, 30(a)	Maths General 2 exams
2015	3	11(d)	14(c)	17, 26(d), 29(b), 30(c)	
2016	14(e)	14(b), 14(d)		8, 27(d), 28(d)	
2017	12(c)	16(b)	15(b)	10, 27(c), 28(c)	
2018	11(d), 14(d)	14(d)	16(c)	19, 26(c), 29(e)	
2019	12(b)	11(d)	16(a)	3, 9, 13, 42	Maths Standard 2 exam
2020 new course	**12**			**26**	

Questions in **bold** can be found in this chapter.

CHAPTER 6
STATISTICS AND
BIVARIATE DATA

STATISTICS AND BIVARIATE DATA 📎

> Other than variance, this entire topic is common content with the Mathematics Standard 2 course.

Types of data

- Categorical: nominal/ordinal
- Numerical: discrete/continuous

Measures of spread

- Range
- Quartiles, deciles and percentiles
- Interquartile range (IQR) $= Q_3 - Q_1$
- Variance, σ^2
- Standard deviation, σ
- Sample standard deviation, s
- An outlier is below $Q_1 - 1.5 \times$ IQR or above $Q_3 + 1.5 \times$ IQR

Measures of central tendency

- Mean, $\bar{x} = \dfrac{\text{sum of values}}{\text{number of values}}$

  $\bar{x} = \dfrac{\Sigma fx}{\Sigma f}$ for fx table

- Median: middle value
- Mode: most common value

The shape of a statistical distribution

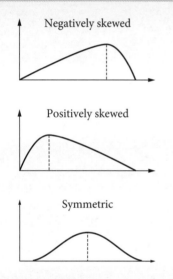

Negatively skewed

Positively skewed

Symmetric

Scatterplots

Used to graph bivariate data (2 variables)

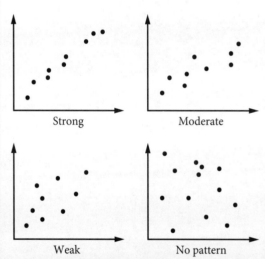

Strong

Moderate

Weak

No pattern

Correlation

- Pearson's correlation coefficient, r, where $-1 \le r \le 1$ for linear relationships:

$r = -1$:	strong, negative
$r = -0.5$:	moderate, negative
$r = 0$:	no correlation
$r = 0.5$:	moderate, positive
$r = 1$:	strong, positive

Line of best fit

- Dependent and independent variables
- Drawing by eye
- Using technology: least-squares regression line
- Interpolation and extrapolation

Glossary

bivariate data
Data relating two variables that have been measured on the same group of people or items, such as height and weight.

categorical data
Data that are names or categories, not numbers; for example, colour.

cumulative frequency
A running total of frequencies of a value in a data set, showing the number of data values below and including that value.

continuous data
Numerical data whose values are measured on a smooth scale (without 'gaps'), such as the response time of a driver.

correlation
The size and direction of a relationship between two or more variables. A correlation between variables, however, does not necessarily mean that one variable *causes* the other variable.

deciles
Nine values that split up a set of ordered data into 10 equal subsections. For example, if a student scores an exam mark placed in the 6th decile, then the mark was better than 60% of students' marks.

dependent variable
A variable with an outcome that depends on another variable, usually *y*, and plotted as *y*-values on a scatterplot.

discrete data
Numerical data whose values are distinct and countable, such as the number of children in a family.

extrapolate
To make predictions using values that are outside the range of the original data. Extrapolation far beyond the range of the original data is not always reliable as it can be misleading.

five-number summary
The minimum value, lower quartile, median, upper quartile and maximum value of a data set.

histogram
A column graph for numerical data, with no gaps between columns.

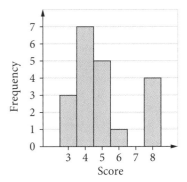

independent variable
A variable with an outcome that does not depend on another variable, usually *x*, and plotted as *x*-values on a scatterplot.

interpolate
To make predictions using values that lie within the range of the original data.

interquartile range
A measure of spread, the difference between the upper and lower quartiles.

least-squares regression line
The **line of best fit** on a scatterplot where the distance of each point from the line is as small as possible, minimising the sum of the squares of the differences. It is the line that best summarises the relationship between two variables.

line of best fit (or regression line)
A line drawn through a scatterplot of data points that most closely represents the relationship between two variables. Can be drawn by eye or by technology that uses a formula.

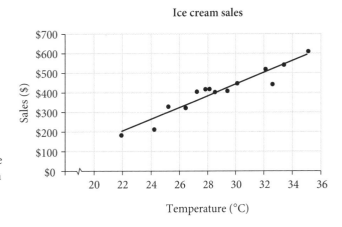

median
The middle data value when the values are placed in order.

measure of central tendency
A summary statistic that describes an entire data set with a single value that represents the middle (or centre) of its distribution. There are three main measures of central tendency: mode, median and mean.

measure of spread

A summary statistic that describes an entire data set with a single value that represents the spread or dispersion of its distribution. Measures of spread include the range, quartiles, interquartile range, variance and standard deviation.

mode

The data value that occurs the most.

negatively skewed

The shape of data where the tail points to the left.

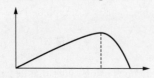

nominal data

Categorical data that cannot be ordered, such as colour.

numerical data

Data that are numbers or values; for example, age.

ordinal data

Categorical data that can be ordered, such as day of the week.

outlier

An extreme or atypical data value that is significantly different from the rest of the data. It is greater than $Q_3 + 1.5 \times IQR$ or less than $Q_1 - 1.5 \times IQR$.

Pareto chart

A graph used in businesses that contains both a bar and a line graph, where individual frequencies are represented by columns in descending order by the bars and the cumulative total is represented by the line graph.

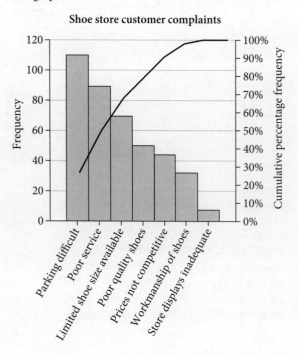

Pearson's correlation coefficient (r)

A statistical value that measures the strength and direction of the linear relationship between a pair of variables. It is represented by r and its value ranges from −1 to 1.

percentile

99 values that split up a set of ordered data into 100 equal subsections. For example, if a student scores an exam mark placed on the 90th percentile, then the mark was better than 90% of students' marks.

population

In statistics, all of the items under investigation from which a sample may be drawn.

positively skewed

The shape of data where the tail points to the right.

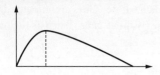

quantile

A value that splits up a set of ordered data into equal subsections. The collective name for quartiles, deciles and percentiles.

quartiles

Three values that split up a set of ordered data into 4 equal subsections. For example, if a student scores an exam mark placed on the 1st quartile, then the mark was better than 25% of students' marks.

range

A measure of spread, the difference between the highest and lowest values in a data set.

scatterplot

A graph of points on a number plane displaying the values of two variables in a bivariate data set.

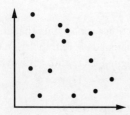

standard deviation

A measure of spread of the data around the mean. It is the square root of the variance. If data values are further from the mean, the higher the standard deviation. If data values are close to the mean, the lower the standard deviation.

variance

A measure of spread of the data around the mean. It is the square of the **standard deviation**.

9780170459228

Topic summary

Descriptive statistics and bivariate data analysis (MA-S2)

> Other than variance, this entire topic is common content with the Mathematics Standard 2 course.

S2.1 Data (grouped and ungrouped) and summary statistics

Types of data

Categorical data (uses words and symbols) and can be:

- **nominal data** (cannot be ordered), such as ice cream flavour

- **ordinal data** (can be ordered), such as size of drink.

Numerical data (uses numbers/quantities) and can be:

- **discrete data** (counted data, distinct, such as whole number values, separate values), such as number of bedrooms

- **continuous data** (measured data, numbers without gaps, such as decimals), such as reaction time.

Example 1

- Favourite pet – dogs, cats, birds, fish – categorical nominal as the list cannot be ordered

- Shoe size – 5, 5.5, 6, 6.5, … – numerical discrete as they are numbers and are separate

- Daily temperature – 20.6°C, 33.2°C – numerical continuous as there are no gaps between values

Graphs and tables

Frequency tables

No. of blueberries in a punnet (x)	No. of punnets (f)
32	12
33	20
34	22
35	18

Histograms

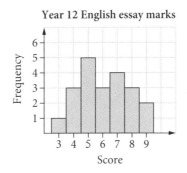

Frequency polygons

Line graph that can be drawn on a histogram, starts and finishes on the horizontal axis, joins the middle of the top of each column.

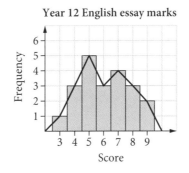

Grouped data and cumulative frequency tables

Grouped data can be sorted into classes with class centres. The cumulative frequency is a running total of the frequencies. This table shows the cumulative frequencies of exam marks.

Scores	Class centre	Frequency	Cumulative frequency
40–49	44.5	1	1
50–59	54.5	4	5
60–69	64.5	8	13
70–79	74.5	6	19
80–89	84.5	5	24
90–99	94.5	1	25

Hint
To find class centre:

$$\frac{40 + 49}{2} = 44.5$$

$$\frac{50 + 59}{2} = 54.5 \quad \text{etc.}$$

Cumulative frequency histograms and polygons (ogives)

This **cumulative frequency** graph shows the number of oranges picked by a group of 45 fruit pickers over summer. 4.5 on the horizontal axis represents the class 0–9.

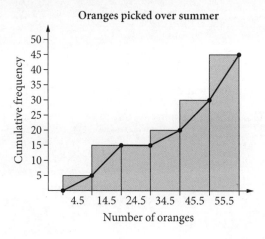

Stem-and-leaf plots and dot plots

Stem-and-leaf plots and dot plots order the data so it can still be seen. (This is a disadvantage of a frequency distribution table.)

Stem	Leaf
4	7
5	1 3 3 8
6	0 2 2 2 5 8 8 9
7	0 1 2 5 5 6
8	3 5 7 8 9
9	4

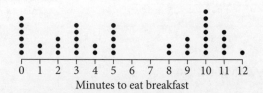

Two-way tables

	Has a dog	Does not have a dog
Has a cat	19	32
Does not have a cat	47	54

Bar charts (for categorical data)

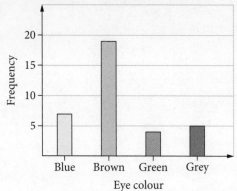

Pie charts or sector graphs

A pie chart is a circle divided into sectors (representing an angle from the centre out of 360° in total).

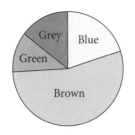

Pareto charts

Pareto charts are often used for quality control in business. Taller columns (left) highlight key issues. The charts combine a bar chart and a line graph.

Individual values are represented in descending order with columns. The cumulative percentage frequency is represented by the line.

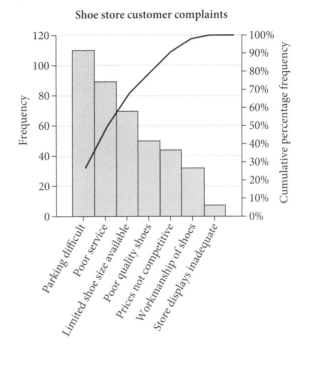

Shoe store customer complaints

Measures of central tendency

$$\text{Mean} = \frac{\text{sum of values}}{\text{number of values}}$$

$$\bar{x} = \frac{\Sigma x}{n}$$

$\bar{x} = \dfrac{\Sigma fx}{\Sigma f}$ for a frequency table.

For the data in the table:

$$\bar{x} = \frac{\Sigma fx}{\Sigma f} = \frac{2422}{72} = 33.638\ldots \approx 33.6.$$

No. of blueberries in a punnet (x)	No. of punnets (f)
32	12
33	20
34	22
35	18

Hint
Σ is the Greek letter 'sigma' meaning 'sum'.

The mean number of blueberries per punnet is approximately 33.6.

The **mode** is the data value that occurs the most. There can be more than one mode in a data set or no mode at all.

The **median** is the middle data value when the values are placed in order.

The median is:

- the middle value for an ODD number of values
- the average of the two middle values for an EVEN number of values.

The median can also be calculated from a cumulative frequency polygon.

Example 2

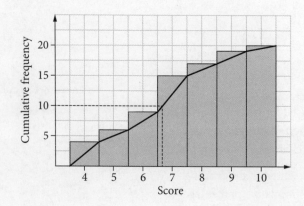

The last column of this cumulative frequency histogram and polygon shows there are 20 values in the data set.

$$20 \div 2 = 10 \text{ gives the position of the median.}$$

On the cumulative frequency polygon, the 10th value is the median (shown by the dotted line).

$$\text{Median} = 7$$

Outliers

An **outlier** is an extreme data value. Some outliers are quite obvious but there is a formal test to check if a value is an outlier:

A data value is an outlier if it is **less than** $Q_1 - 1.5 \times IQR$ or **greater than** $Q_3 + 1.5 \times IQR$.

Measure of central tendency	Features	When is it most appropriate?
Mean $\bar{x} = \dfrac{\text{sum of values}}{\text{number of values}}$ $= \dfrac{\Sigma x}{n}$ $= \dfrac{\Sigma fx}{\Sigma x}$	• Depends on all values in the data set • Is affected by outliers	When the data set does not have many outliers
Median Middle value or average of two middle values	• Not affected by outliers	When the data set has many outliers; for example, house prices, salaries
Mode Most popular value(s)	• Not affected by outliers	When the most common value or category is needed (for example, dress size); also useful for categorical data

Quartiles, deciles and percentiles

The general name for quartiles, deciles and percentiles is **quantiles**.

Quartiles

A **quartile** divides data into quarters. Commonly seen in a box plot.

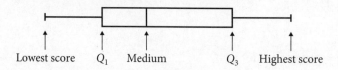

The median, lower and upper quartiles can be read off a cumulative frequency polygon.

Example 3

As there are 20 values, 20 ÷ 2 = 10th value (as in Example 2).

Reading from 10 on the cumulative frequency axis,

median is 7.

The lower quartile (Q_1) = 20 ÷ 4 = 5th value

Reading from 5 on the cumulative frequency axis:

$$Q_1 = 5$$

The upper quartile (Q_3) = $20 \times \dfrac{3}{4}$

$$= 15$$

Reading from 15 on the cumulative frequency axis:

$$Q_3 = 7.5$$

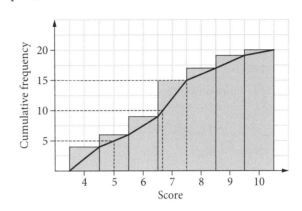

Deciles and percentiles

A **decile** divides the data set into 10 equal parts.

A **percentile** divides a data set into 100 equal parts.

Example 4

The cumulative frequency histogram and polygon show the number of hours a group spends rehearsing each week over 25 weeks, preparing for the opening of a live theatre production.

Find the 35th percentile and 7th decile.

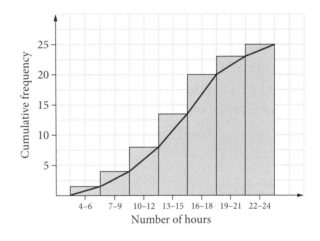

Solution

Note: Class centres are calculated for the number of hours on the horizontal axis below.

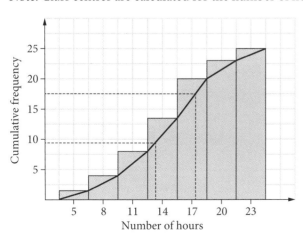

The cumulative frequency is 25.

Find the 35th percentile:

$$0.35 \times 25 = 8.75\text{th score}$$

From the graph, the 35th percentile is 13. (This means 35% of rehearsals were less than 13 hours).

Find the 7th decile:

$$0.7 \times 25 = 17.5\text{th score}$$

The 7th decile is 17. (This means 70% of rehearsals were less than 17 hours).

Measures of spread

Range = highest value – lowest value

The range is affected by outliers.

Interquartile range (IQR) = upper quartile (Q_3) – lower quartile (Q_1)

The IQR represents the middle 50% of the data, so is unaffected by outliers.

Example 5

Find the range and interquartile range (IQR) of these scores:

$$8 \quad 12 \quad 6 \quad 14 \quad 20 \quad 23 \quad 18 \quad 17 \quad 9$$

Solution

Range = highest score – lowest score = 23 – 6 = 17

To find the IQR, list the 9 scores in ascending order:

$$6 \quad 8 \mid 9 \quad 12 \quad \mathbf{14} \quad 17 \quad 18 \mid 20 \quad 23$$
$$ Q_1 \qquad \text{Median} \qquad Q_3$$

$$Q_1 = \frac{8+9}{2} = 8.5$$

$$Q_3 = \frac{18+20}{2} = 19$$

IQR = 19 – 8.5 = 8.5

Variance measures how far the values in a data set are from the mean and, like the mean, every value in the data set is used to calculate it. You do not need to learn the formula for variance because it can be calculated using the calculator's statistics mode (see Example 6 below).

Standard deviation is another **measure of spread**, and it is simply the square root of the variance. The symbol for standard deviation is σ and the symbol for variance is σ^2.

Variance and standard deviation were also covered with discrete probability distributions in Year 11.

Example 6

Here are the instructions for calculating the mean, standard deviation and variance of this data set.

$$30 \quad 28 \quad 26 \quad 31 \quad 34 \quad 35 \quad 32 \quad 33 \quad 21 \quad 25 \quad 28 \quad 32 \quad 32 \quad 35$$

Operation	Casio Scientific	Sharp Scientific
Start statistics mode.	**MODE** STAT 1-VAR	**MODE** STAT **=**
Clear the statistical memory.	**SHIFT** 1 Edit, Del-A	**2ndF** **DEL**
Enter data.	**SHIFT** 1 Data to get table 30 **=** 28 **=**, etc. to enter in column **AC** to leave table	30 **M+** 28 **M+**, etc.
Calculate the mean. ($\bar{x} = 30.14285\ldots$)	**SHIFT** 1 Var $\bar{x}$ **=**	**RCL** $\bar{x}$
Calculate the standard deviation ($\sigma_x = 3.9434\ldots$)	**SHIFT** 1 Var σ_x **=**	**RCL** σx
Check the number of scores, ($n = 14$)	**SHIFT** 1 Var n **=**	**RCL** n
Return to normal (COMP) mode.	**MODE** COMP	**MODE** 0

For the variance, simply square the standard deviation:

$$\sigma_x^2 = (3.9434\ldots)^2 = 15.5510\ldots$$

Note: When calculating the standard deviation of a set of data, we usually use the population standard deviation, σ. If the set of data is a sample, however, then we can use the sample standard deviation, s, to estimate the results for a **population**. For the above data:

$$s_x = 4.0923\ldots$$

Shape and modality

The shape of data also helps us make sense of measures of central tendency and spread.

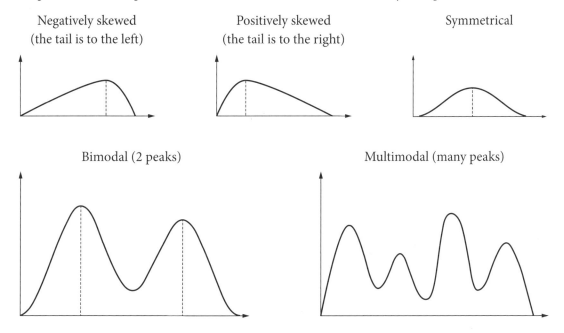

Negatively skewed
(the tail is to the left)

Positively skewed
(the tail is to the right)

Symmetrical

Bimodal (2 peaks)

Multimodal (many peaks)

S2.2 Bivariate data analysis

Bivariate data measures 2 variables on the same data set to see if they correlate with (are related to) each other; for example, the age and height of a group of athletes.

Scatterplots are used to graph bivariate data.

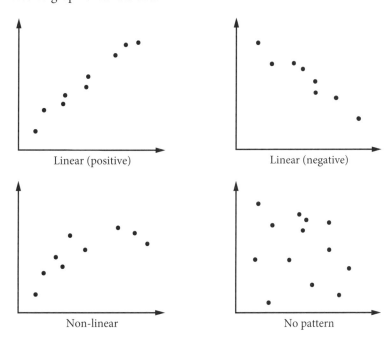

Linear (positive)

Linear (negative)

Non-linear

No pattern

Correlation measures how well 2 variables are related if there appears to be a linear relationship between them.

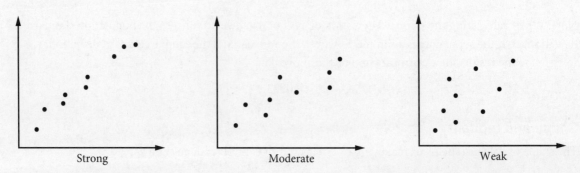

Strong   Moderate   Weak

Pearson's correlation coefficient (r)

We can measure the correlation coefficient, r, for linear relationships, $-1 \leq r \leq 1$.

$r = -1$:	strong, negative
$r = -0.5$:	moderate, negative
$r = 0$:	no correlation
$r = 0.5$:	moderate, positive
$r = 1$:	strong, positive

> **Hint**
> Strength and direction *must* be used to define the correlation between 2 variables.

The correlation coefficient can be calculated using the calculator's STAT mode.

Example 7

A group of students were surveyed about the number of hours they studied and their results in a Science exam.

Find Pearson's correlation coefficient for this bivariate data, correct to three decimal places.

No. of hours studying	7	5	10	8	12	9	12
Exam result (%)	72	48	75	65	73	81	84

Solution

Operation	Casio Scientific	Sharp Scientific
Place your calculator in statistical mode.	MODE 2 : STAT 2 : A+BX	MODE 1 STAT 1 LINE
Clear the statistical memory.	SHIFT 1 3 : Edit 2 : Del-A	2ndF DEL
Enter data.	SHIFT 1 2 : Data 7 ▣ 5 ▣ etc. for 1st column 72 ▣ 48 ▣ etc. for 2nd column AC	7 2ndF STO 72 M+ 5 2ndF STO 48 M+ etc.
Calculate r.	SHIFT 1 5 : Reg 3 : r ▣	ALPHA r ▣
Change back to normal mode.	MODE 1 : COMP	MODE 0

$r = 0.784\,135\ldots$

≈ 0.784

9780170459228

Line of best fit

We can draw a **line of best fit** to identify and summarise trends in scatterplots. We draw a regression line if there is a linear correlation and find its equation. This helps wus make predictions using the data. A line of best fit can be hand drawn 'by eye' or calculated more precisely using a calculator and formulas.

For a line of best fit 'by eye', use a ruler and draw a line that represents as many data points as possible. Draw a line with roughly half the points above it and half below it.

Example 8

An ice cream stand keeps a record over a fortnight of how much ice cream is sold each day (in dollars) and the temperature at midday each day. Their figures are shown in a table on the right.

Temperature °C	Ice cream sales
24.2	$215
26.4	$325
21.9	$185
25.2	$332
28.5	$406
32.1	$522
29.4	$412
35.1	$614
33.4	$544
28.1	$421
32.6	$445
27.2	$408
30.1	$450
27.8	$420

The data is plotted on the axes shown below.

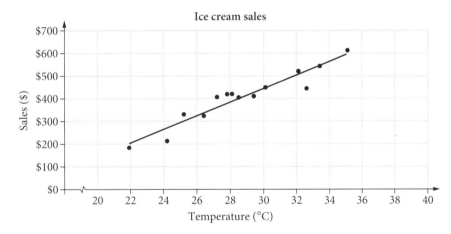

The line of best fit is added and its equation can be found.

Choose 2 points on the line, such as $(33.4, 544)$ and $(30.1, 450)$.

Gradient:

$$m = \frac{y_2 - y_1}{x_2 - x_1}$$

$$= \frac{544 - 450}{33.4 - 30.1}$$

$$= \frac{94}{3.3}$$

$$= 28.4848...$$

$$\approx 28.5$$

Equation: using $m = 28.5$ and point $(30.1, 450)$

$$y - y_1 = m(x - x_1)$$

$$y - 450 = 28.5(x - 30.1)$$

$$= 28.5x - 857.85$$

$$y = 28.5x - 407.85$$

> **Hint**
>
> Because y depends on x, y is called the **dependent variable** and x the **independent variable**. The dependent variable is always graphed on the vertical axis.

TOPIC SUMMARY

We can use the trendline to make predictions.

Interpolation uses the model to make predictions about data values lying *within* the range of the data provided.

For example, when the temperature is 31°C, the sales could be approximately $475.

Extrapolation uses the model to make predictions *beyond* the range of the data provided.

For example, when the temperature is 38°C, the sales could be approximately valued at $690.

However, it may not be a good idea to extrapolate, as this line may only be accurate for the known data, and not for values much higher or lower than the known data.

Least-squares regression line

The most popular and accurate line of best fit is the **least-squares regression line**, where the squares of the distances from each point in the scatterplot to the line are minimised, as shown in the diagram.

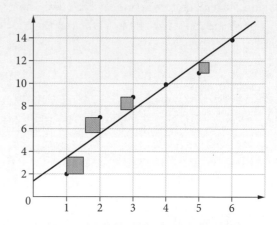

As with standard deviation and correlation, the equation of the least-squares regression line is calculated using technology, such as a scientific calculator, spreadsheet or dynamic geometry software/website. For a scientific calculator, see Example 7 on page 144 for the calculator steps required to calculate Pearson's correlation coefficient for study hours vs exam results, but change the last step to the following:

Operation	Casio Scientific	Sharp Scientific
On these calculators, the gradient is b and the y-intercept is a ($b = 3.625$, $a = 38.5178…$).	SHIFT 1 Reg b = SHIFT 1 Reg a =	ALPHA b = ALPHA a =

The calculator's equation for the least-squares line is $y = bx + a$, so it is $y = 3.625x + 38.518$.

Practice set 1

Multiple-choice questions

Solutions start on page 160.

Question 1

The histogram below shows the distribution of the population sizes of 35 countries in 2020.

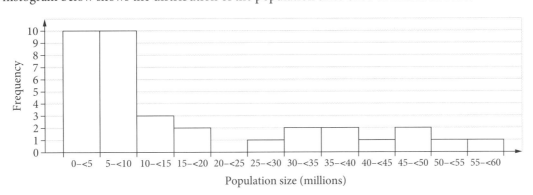

How many countries had a population size of between 5 and 30 million?

A 6   **B** 8   **C** 16   **D** 18

Question 2

What is the shape of the distribution shown in Question **1**?

A Unimodal   **B** Symmetric

C Negatively skewed   **D** Positively skewed

Question 3

The stem-and-leaf plot below displays the average CO_2 emissions per person (in tonnes) from 31 countries.

Stem	Leaf
0	1 2 3 3 4 5 6 6 7
1	0 1 3 6 9
2	0 4 4 7 8 9
3	6 8 8
4	1 3 8
5	1 2 7
6	
7	0 3

Key: $0|2 = 0.2$

Which of the following best describes this data?

A Negatively skewed with a median of 24 tonnes and a range of 72

B Positively skewed with a median of 24 tonnes and a range of 72

C Negatively skewed with a median of 2.4 tonnes and a range of 7.2

D Positively skewed with a median of 2.4 tonnes and a range of 7.2

Question 4

What type of data is the volume of water in a dam?

A Categorical nominal   **B** Categorical ordinal

C Numerical discrete   **D** Numerical continuous

Question 5 〇〇●

This Pareto chart shows complaints about holiday apartment rentals.

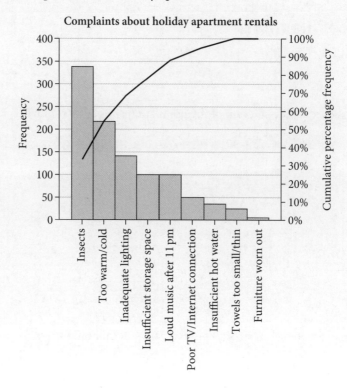

Which one of the following statements is false?

A Most holiday renters are satisfied with the quality of the TV/Internet connection.

B Worn out furniture is a major concern of holiday renters.

C Holiday renters are more concerned about the heating/cooling of the apartments than they are about inadequate lighting and insufficient hot water combined.

D 25% of holiday makers complained about loud music after 11 pm.

Question 6 〇〇●

On a reality TV show, viewers voted online for their favourite celebrity. The results are shown in the table below.

Celebrity	Male voters	Female voters
Jayden	1276	4163
Bree	2104	3526
Suki	1532	3879

One female voter was selected at random from all the female voters.

What is the probability that she voted for Jayden?

A $\dfrac{4163}{16\,480}$ **B** $\dfrac{4163}{11\,568}$ **C** $\dfrac{4163}{7405}$ **D** $\dfrac{4163}{5439}$

Question 7 ●●▨

The graph below shows the monthly rainfall for Moss Vale in millimetres over a year.

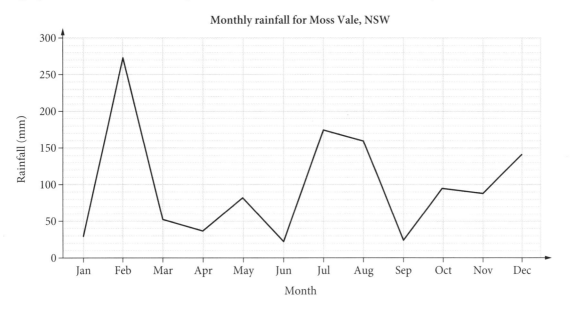

What was the median monthly rainfall for Moss Vale in the year the data was collected?

A 65   **B** 87   **C** 98   **D** 125

Question 8 ●▨▨

This stem-and-leaf plot shows the distribution of history test scores for a class of 19 students.

What is the range of test scores?

A 47   **B** 58

C 64   **D** 66

Stem	Leaf
4	0 1 7
5	3 6 8 8 8
6	1 4 5 6 6 9
7	0 0 5
8	2 7

Key: 4|0 = 40

Question 9 ●●▨

What is the interquartile range of the test scores in Question 8?

A 11   **B** 12   **C** 14   **D** 17

Question 10 ●●▨

A set of Geography yearly exam results are shown in a cumulative frequency histogram and polygon below.

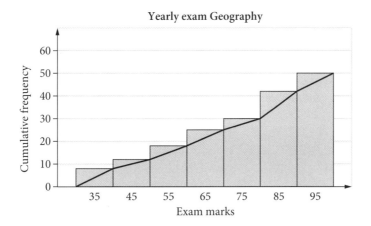

Max knows that his exam result was in the 6th decile. What was Max's result?

A 66   **B** 72   **C** 80   **D** 85

Question 11 ©NESA 2013 HSC EXAM, QUESTION 6 ●●

A survey was conducted where people were asked which of two brands of smartphones they preferred. The results were:

- 48% preferred Brand X

- 52% preferred Brand Y

A graph displaying the data is to be included in a magazine article. The editor of the magazine wishes to ensure that the graph is not misleading in any way.

Which graph should the editor choose to include in the article?

A

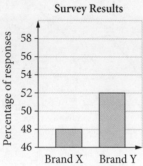

B

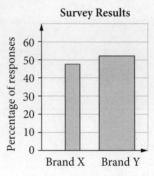

C

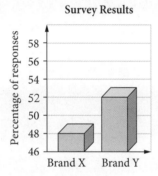

D

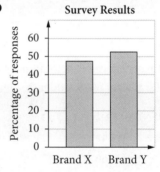

Question 12 ●●

The mean of a data set is 12 and the standard deviation is 4.5.

Hamish increases each value in the data set by 5.

Which of the following statements will be true?

A The mean and standard deviation will both increase by 5.

B The mean will increase by 5 and the standard deviation will stay the same.

C The mean will stay the same and the standard deviation will increase by 5.

D The mean and standard deviation will both stay the same.

Question 13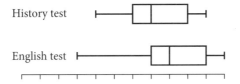

The marks for a History test and an English test are presented in box plots.

History test

English test

Which statement is correct?

A The marks for the History test have a higher interquartile range than marks for the English test.

B The median for History test marks is higher than the median for English test marks.

C The box-and-whisker plots for both History and English show a negative skew.

D 75% of the English test marks are higher than 50% of the History test marks.

Question 14

Which of the following questions is the most clear and appropriate to include in a survey (not misleading or biased)?

A How much can you save by shopping at factory outlets?

B What is your favourite movie?

C On how many days in the last week did you catch public transport?

D Do you play cricket or tennis?

Question 15

For the data shown in the scatterplot, which value is most likely to be Pearson's correlation coefficient?

A −0.86

B −0.58

C 0.16

D 0.72

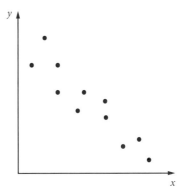

Question 16

An environmental scientist measures the mean mass, m kg, of certain species of fish against the mean concentration, in C mg/L, of a chemical in a river system.

The results are shown in the table below.

m (kg)	6.5	7.2	7.4	7.6	8.3	9.7
C (mg/L)	1.94	1.78	1.62	1.51	1.52	1.4

Which of the following is the equation of the least-squares regression line?

A $C = 2.848 - 0.157m$

B $C = 15.65 - 4.832m$

C $C = 2.99 - 0.168m$

D $C = 4.106 - 0.323m$

Question 17 ●●

The ages of 11 adults attending a dance class are:

$$55 \quad 21 \quad 20 \quad 26 \quad 23 \quad 28 \quad 35 \quad 19 \quad 67 \quad 23 \quad 49$$

Which one of the following statements is false?

A The median age is 26.

B The lower quartile is 21.

C The IQR is 29.

D The person aged 67 is not an outlier.

Question 18 ●●

Which of the these relationships is most likely to have a negative correlation?

A The amount of rainfall and the number of umbrellas sold

B The number of pets per household and the number of mobile phones per household

C The hours spent training for a swimming race and the time taken to complete the swimming race

D The hours spent in direct sunlight and the chance of sunburn

Question 19 ●●

A sample of 16 people were asked to indicate the time (in hours) they had spent on their mobile phones the previous night. The results are displayed in the dot plot on the right.

What is the mean and sample standard deviation of these times?

A $\bar{x} = 2.63, s = 2.00$

B $\bar{x} = 2.63, s = 1.93$

C $\bar{x} = 2.33, s = 1.62$

D $\bar{x} = 2.33, s = 1.68$

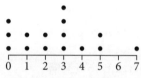
Time spent on mobile phone

Question 20 ●●●

The table below shows the hourly rate of pay paid to 10 employees in a company in 1999 and 2019.

Find the correlation coefficient, r, for this set of data.

A 0.81

B 0.85

C 0.92

D 0.98

Employee	Hourly rate of pay ($)	
	1999	2019
Jack	9.67	15.03
Tracey	13.03	20.52
Claire	19.30	29.10
Erin	9.90	16.12
Karis	12.14	19.70
Arjun	8.85	15.55
Hamish	8.97	17.95
Jung Hee	11.04	18.37
Satara	19.06	27.80
Ali	13.67	21.60

Practice set 2

Short-answer questions

Solutions start on page 162.

Question 1 (5 marks)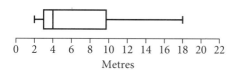

The heights of trees in North Park were measured and the data represented in the box plot below.

Heights of trees in North Park

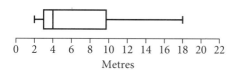

Metres

a State the five-number summary for this data.   2 marks

b Calculate the interquartile range.   1 mark

c Describe the meaning of the interquartile range.   1 mark

d Describe the shape of the data.   1 mark

Question 2 (4 marks)

The frequency histogram shows the results of a survey of households and the number of pets in each household.

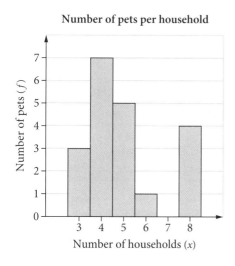

Number of pets per household

a How many households were surveyed in total?   1 mark

b Complete the frequency distribution table below.   2 marks

No. of households (x)	No. of pets (f)	$f \times x$
3		
4		
5		
6		
7		
8		

c Hence, calculate the mean number of pets per household.   1 mark

Question 3 (3 marks)

The parallel box plots show the heights of students in Year 10 and Year 9.

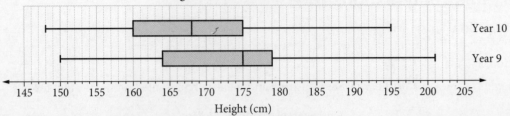

Heights of Year 9 and 10 students

a What percentage of students in Year 10 are taller than 175 cm?   1 mark

b The number of students taller than 175 cm is the same for both the Year 10 and   2 marks
Year 9 groups. If Year 10 has 220 students, how many more students are there in
Year 10 than in Year 9?

Question 4 (6 marks)

The ages of 9 employees in an office are shown.

$$41 \quad 27 \quad 19 \quad 25 \quad 23 \quad 28 \quad 36 \quad 27 \quad 68$$

a Are there any outliers? Show working to justify your answer.   2 marks

b Calculate the mean (answer to one decimal place) and the median.   2 marks

c Which is the better measure of central tendency? Refer to calculations to justify   2 marks
your answer.

Question 5 (4 marks)

A group of 150 students were surveyed and the results recorded.

	Likes cricket	**Does not like cricket**	**Total**
Female	38	21	A
Male	65	26	91
	103	B	150

a Calculate the values of A and B.   2 marks

b What percentage of those who don't like cricket are female? Answer correct to   2 marks
one decimal place.

Question 6 (2 marks)

A cumulative frequency table for a set of data is shown below.

Score	Cumulative frequency
1	5
2	9
3	16
4	20
5	34
6	42

Calculate the interquartile range for this set of data.   2 marks

Question 7 (4 marks)

A class of 30 students sat a Geography test and a History test. The results are shown in a scatterplot with a line of best fit.

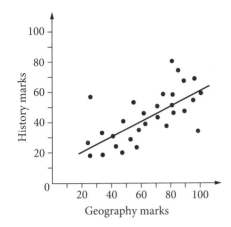

a How many students scored less than 40 in both the Geography and the History tests? 1 mark

b Describe the correlation between the Geography and History test results. 1 mark

c Examine the scatterplot and comment on the following statement.

In this class all students who scored very high marks in Geography also scored very high marks in History.

Justify your answer. 2 marks

Question 8 (4 marks) ©NESA 2013 GENERAL MATHEMATICS HSC EXAM, QUESTION 26(f)

Jason travels to work by car on all five days of the working week, leaving home at 7 am each day. He compares his travel times using roads without tolls and roads with tolls, over a period of 12 working weeks.

He records his travel times (in minutes) in a back-to-back stem-and-leaf plot.

Travel time (minutes)

Without tolls		With tolls
9	3	5 8 9 9
9 9 8 7 7 6 5 5 4 4 3 2 0	4	0 1 2 6 7 7 8 8 8 9
9 8 7 5 4 3 3 3 2 2 2 2 1 1 0	5	2 4 4 5 6 8 9
1	6	1 3 5 7
	7	0 2 8
	8	2
	9	0

a What is the modal travel time when he uses roads without tolls? 1 mark

b What is the median travel time when he uses roads without tolls? 1 mark

c Describe how the two data sets differ in terms of the spread and skewness of their distributions. 2 marks

Question 9 (8 marks)

The final jumps recorded for two teenage athletes in 8 long jump sessions are shown in the table.

Ben	1.80 m	1.80 m	1.90 m	1.70 m	1.80 m	2.00 m	1.60 m	1.90 m
Lamar	2.00 m	1.90 m	1.70 m	2.10 m	1.90 m	2.10 m	2.10 m	1.80 m

a Calculate the median long jump distance for: 2 marks

  i Ben

  ii Lamar

b Calculate the interquartile range for: 2 marks

  i Ben

  ii Lamar

c State the five-number summary for: 2 marks

  i Ben

  ii Lamar

d Hence, use an appropriate scale (in metres) to draw parallel box-and-whisker plots for each athlete's recorded long jumps.   2 marks

Question 10 (4 marks)

Mackenzie called her internet provider on 6 occasions to report problems with internet connection. On each occasion she wrote down the length of time she waited to speak to a customer service operator. The times (in minutes) were:

$$16 \quad 4 \quad 5 \quad 20 \quad 14 \quad 12$$

a Calculate the mean and standard deviation of these times, correct to one decimal place.   2 marks

b Zala also rang the same internet provider, several times, to report connection problems. Her mean waiting time was 15 minutes and the standard deviation was 4.3 minutes. Compare Mackenzie's waiting times with Zala's. Write 2 valid observations with reference to your calculations.   2 marks

Question 11 (6 marks)

This table lists the number of permanent arrivals to Australia in April each year from 2011 to 2020. The data for 2015 is missing.

Year	2011	2012	2013	2014	2015	2016	2017	2018	2019	2020
Number of arrivals	11 210	13 340	12 820	10 800		11 310	10 820	9390	7780	410

Source: www.abs.gov.au

Use the table to answer the following questions.

a Draw a scatterplot of the data. (For the vertical axis, use 0, 2000, 4000, 6000, …, 14 000.)   2 marks

b Draw a line of best fit on your scatterplot.   1 mark

c Use interpolation to estimate the number of permanent arrivals to Australia in April 2015.   1 mark

d In late March 2020, Australia closed its borders. Why did this occur?   1 mark

e Explain why extrapolation does not provide the best method for predicting the number of permanent arrivals in April 2021.   1 mark

Question 12 (1 mark)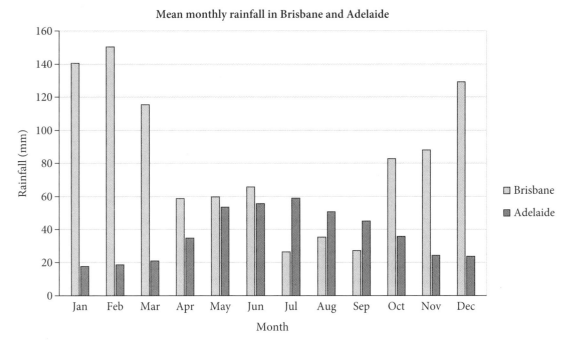

The graph shows the mean monthly rainfall in Brisbane and Adelaide over 12 months.

Which city had the smaller spread of rainfall over the 12 months? Without referring to any   1 mark
calculations, justify your answer.

Question 13 (4 marks)

The table below shows the English marks (x) and the Science marks (y) for a class of 12 students.
The Science mark for student M is missing.

	A	B	C	D	E	F	G	H	J	K	L	M
x	67	65	61	75	67	75	85	87	69	89	85	80
y	56	66	69	65	68	72	76	84	72	82	80	

a   Calculate the correlation coefficient for this data, without including M, and describe   2 marks
the correlation. Give your answer to two decimal places.

b   Find the equation of the least-squares regression line. Hence, estimate the Science mark   2 marks
for student M, correct to the nearest whole number.

Question 14 (4 marks)

The test scores for 7 students are given:

$$5 \quad 5 \quad 7 \quad 6 \quad 6 \quad x \quad 4 \quad 9$$

a   Find the lowest possible test score for x given that the median is 6.   1 mark

b   Find a possible value of x given that the variance is 2.   3 marks

Question 15 (4 marks) ⚫⚫⬜

Data for life expectancy (expected remaining years of life) for males at selected ages are given in the table.

Age (x years)	Life expectancy for males (y years)
0	80.9
5	76.2
10	71.2
15	66.3
20	61.4
25	56.6
30	51.8
35	47.0
40	42.2

a   Find the equation of the least-squares regression line in the form $y = mx + c$, where m and c are correct to two decimal places.   1 mark

b   Dom uses the least-squares regression line from part **a** to estimate the life expectancy of his father, who is currently 79. Explain why this is NOT a valid estimate.   2 marks

c   Dom is a male aged 36. Use the least-squares regression equation to calculate his life expectancy.   1 mark

Question 16 (2 marks) ⚫⚫⚫

There were 6 members of a basketball team who had a mean individual score of 11 points per game. Sarah joined the team and the mean points score increased to 12 points per game.

What is Sarah's individual score?   2 marks

Question 17 (4 marks) ⚫⚫⚫

The results of a survey of a group of Year 12 students and the amount of time in a week typically spent on 2 social media sites were recorded and displayed in the box-and-whisker plots shown below. An equal number of Year 12 students used Chatter and InstaTik.

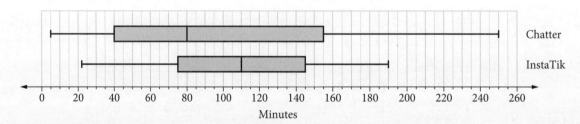

a   What is the interquartile range for Chatter?   1 mark

b   Compare the two data sets by referring to the skewness of the distributions, the measures of central tendency and measures of spread.   3 marks

Question 18 (3 marks) ⚫⚫⬜

Five students sat both a Drama and a Biology exam. Their results are shown in the table.

Drama	75	55	81	64	72
Biology	84	48	77	53	68

The correlation coefficient for this data is approximately 0.9.

a   Find the value of the correlation coefficient using your calculator, correct to three decimal places.   1 mark

b   Find the equation of the least-squares line of best fit for this data, writing the gradient and y-intercept, correct to two decimal places.   2 marks

Question 19 (3 marks) ⬤⬤⬤

A study was conducted to investigate the effect of energy drinks on sleep. In this study, the number of cans of energy drinks consumed and the amount of sleep (in hours) were recorded for a group of adults.

The summary statistics recorded are shown below.

	Energy drink (cans)	Sleep (hours)
Mean	2.42	7.08
Standard deviation	1.56	1.12
Correlation coefficient (r)	−0.770	

The equation of the least-squares line of best fit is $y = mx + c$, where:

$$m = r \times \frac{\text{standard deviation of } y\text{-scores}}{\text{standard deviation of } x\text{-scores}} \left(\text{or } \frac{r\sigma_y}{\sigma_x} \right)$$

$$c = \text{mean of } y\text{-scores} - m \times \text{mean of } x\text{-scores} = \bar{y} - m\bar{x}$$

a Describe the relationship between the number of energy drinks consumed and the amount of sleep taken.   1 mark

b For every additional can of energy drink consumed, calculate by how much the amount of sleep changes.   2 marks

Question 20 (5 marks) ©NESA 2020 HSC EXAM, QUESTION 27 ⬤⬤⬤

A cricket is an insect. The male cricket produces a chirping sound.

A scientist wants to explore the relationship between the temperature in degrees Celsius and the number of cricket chirps heard in a 15-second time interval.

Once a day for 20 days, the scientist collects data. Based on the 20 data points, the scientist provides the information below.

A box plot of the temperature data is shown.

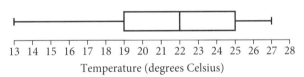

Temperature (degrees Celsius)

- The mean temperature in the dataset is 0.525°C below the median temperature in the dataset.

- A total of 684 chirps was counted when collecting the 20 data points.

The scientist fits a least-squares regression line using the data (x, y), where x is the temperature in degrees Celsius and y is the number of chirps heard in a 15-second time interval. The equation of this line is

$$y = -10.6063 + bx,$$

where b is the slope of the regression line.

The least-squares regression line passes through the point $(\bar{x}, \bar{y})$, where $\bar{x}$ is the sample mean of the temperature data and $\bar{y}$ is the sample mean of the chirp data.

Calculate the number of chirps expected in a 15-second interval when the temperature is 19° Celsius. Give your answer correct to the nearest whole number.   5 marks

Practice set 1

Worked solutions

1 C

$10 + 3 + 2 + 1 = 16$

2 D

It is positively skewed.

3 D

Tail points to higher values, so positively skewed.

Median (16th value) = 2.4
Range = $7.3 - 0.1 = 7.2$

4 D

Numerical continuous

5 B

False, worn out furniture is the lowest (< 5%).

6 B

Total female voters = 11 568

Females voting for Jayden = 4163

Probability = $\dfrac{4163}{11\,568}$

7 B

June	Sept	Jan	Apr	Mar	May	Oct
22	23	31	38	51	80	96

Graphically, the median would be closest to 87.

8 A

Range = $87 - 40 = 47$

9 C

$19 \div 2 = 9.5$, need 10th value, so median = 64

For lower quartile:
40 41 47 53 **56** 58 58 58 61 | 64

$Q_1 = 56$

For upper quartile:
64 | 65 66 66 69 **70** 70 75 82 87

$Q_3 = 70$

IQR = $Q_3 - Q_1 = 70 - 56 = 14$

10 C

For 6th decile: $0.6 \times 50 = 30$

Read 30 vertically from graph: 80

11 D

A: No; scale should start at 0

B: No; columns are of different widths, difficult to compare heights

C: No; 3D so Brand Y looks 3 times the volume of Brand X, poor vertical scale (hard to read accurately from 3D, scale does not start at 0)

D: Correct; good scale, starting at 0 and equal width of columns

12 B

The standard deviation will not change as the data values will be the same distance apart. The mean will increase by 5.

13 D

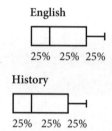

14 C

A: No, too open-ended and savings could be considered different to various people

B: No, potentially creates a very large list of movies

C: Yes, quantitative, definite time of 1 week is easier to document

D: No, might play neither; too specific to provide 2 sports to choose from

15 A

Strong negative correlation

16 A

Using STAT mode on a calculator and 'regression' option

$C = 2.848 - 0.157m$

17 C

19 20 **21** 23 23 **26** 28 35 **49** 55 67

Median = 26

$Q_1 = 21$

$Q_3 = 49$

IQR = 49 − 21 = 28 (not 29)

Outlier > Q_3 + 1.5 × IQR = 49 + 1.5 × 28 = 91

Outlier must be greater than 91

67 < 91, so not an outlier

18 C

More swimming training should result in lower race times.

19 A

Use STAT mode on a calculator:

x	f
0	3
1	2
2	2
3	5
4	1
5	2
6	0
7	1

$\bar{x} = 2.63$
$s = 2.00$

20 D

Use STAT mode on a calculator and 'regression': $r = 0.98$

Practice set 2

Worked solutions

Question 1

a

Min	2
Q_1	3
Median	4
Q_3	9.5
Max	18

b $IQR = Q_3 - Q_1 = 9.5 - 3 = 6.5$

c 50% of the trees in North Park are between 3 and 9.5 metres tall.

d The data is positively skewed.

Question 2

a $3 + 7 + 5 + 1 + 4 = 20$

b

x	f	fx
3	3	9
4	7	28
5	5	25
6	1	6
7	0	0
8	4	32
Total	20	100

c $\bar{x} = \dfrac{100}{20} = 5$ pets

Question 3

a 25%

b Year 10: 25% × 220 = 55 students

Year 9:   50% = 55

100% = 55 × 2 = 110 students

220 − 110 = 110 more students in Year 10 than in Year 9

Question 4

a In ascending order:

19 23 25 27 **27** 28 36 41 68

Median = 27

$Q_1 = \dfrac{23 + 25}{2} = 24$

$Q_3 = \dfrac{36 + 41}{2} = 38.5$

$IQR = Q_3 - Q_1 = 38.5 - 24 = 14.5$

$Q_1 - 1.5 \times IQR = 24 - 1.5 \times 14.5 = 2.25$

$19 > 2.25$

19 is not an outlier.

$Q_3 + 1.5 \times IQR = 38.5 + 1.5 \times 14.5 = 60.25$

$68 > 60.25$

1 outlier = 68

b Mean = 294 ÷ 9 = 32.7

Median = 27

c The median (27) is better measure of central tendency as it is unaffected by the outlier (68), whereas the mean (32.7) is.

Question 5

a $A = 38 + 21 = 59$

$B = 21 + 26 = 47$

b $\dfrac{21}{47} \times 100\% = 44.680\ldots$
$\approx 44.7\%$

Question 6

$42 \div 2 = 21$

To find the median, add and average the 21st, therefore the 22nd values.

Median $= \dfrac{5 + 5}{2} = 5$

$0.25 \times 42 = 10.5$

$Q_1 = 3$

$0.75 \times 42 = 31.5$

$Q_3 = 5$

$IQR = 5 - 3 = 2$

Question 7

a 4

b Moderate, positive linear relationship

c Although this is a true statement for most of the high Geography and high History marks, there is one student with an almost perfect score in Geography but less than 35 in History, which does not support this statement.

Question 8

a Mode = 52 min

b Median $= \dfrac{50 + 51}{2} = 50.5$ min

c The 'without tolls' data has less spread and is more bunched, is symmetric and data is clustered in the 40s and 50s. The 'with tolls' data is positively skewed with clusters in the 40s and 50s and has a greater spread of times: 30s–90s.

Question 9

a i Ben: (ascending order, m)

   1.60 1.70 | 1.80 1.80 | 1.80 1.90 | 1.90 2.00
   $\quad Q_1 \qquad\quad$ median $\qquad\quad Q_3$

   Median = 1.80 m

ii Lamar (descending order):

   1.70 1.80 | 1.90 1.90 | 2.00 2.10 | 2.10 2.10
   $\quad Q_1 \qquad\quad$ median $\qquad\quad Q_3$

   Median = 1.95 m

b i Ben: IQR = 1.90 − 1.75 = 0.15 m

ii Lamar: IQR = 2.10 − 1.85 = 0.25 m

c i Ben: 1.60, 1.75, 1.80, 1.90, 2.00

ii Lamar: 1.70, 1.85, 1.95, 2.10, 2.10

d

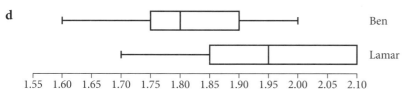

Question 10

a Mackenzie: mean = 11.8 minutes, standard deviation = 5.7 minutes

b On average Zala's waiting time was longer. Zala's waiting times were, however, more consistent as the standard deviation was smaller than Mackenzie's standard deviation.

Question 11

a and b

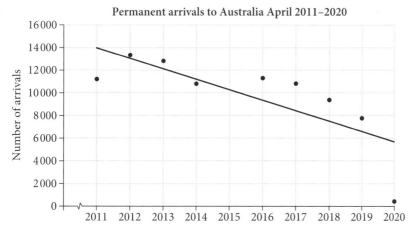

c 10 200 (approx.)

d Borders were closed due to a world pandemic (COVID-19).

e The pandemic may be over due to a vaccine having been developed and the borders reopened, therefore the numbers of permanent arrivals should not be as low as in 2020; possible return to pre-pandemic levels (or similar).

Question 12

The spread for Adelaide is smaller as there is less variation (cluster May to August) in the mean monthly rainfall than for Brisbane, which has a higher mean rainfall in 2 clusters, January to March and October to December.

Question 13

a $r = 0.80$

Strong, positive correlation

b $y = 21.9 + 0.67x$

$x = 80, y = 21.9 + 0.67 \times 80 = 75.1$

Science mark for student M = 75

Question 14

a Rearrange data in order:
4 5 5 6 6 7 9 x
For median to be 6, x has to be at least 6.

b By trial-and-error, using the STAT mode of the calculator, $x = 6$.

OR

$$\sigma^2 = \frac{\Sigma x^2}{8} - \bar{x}^2 = 2$$

$$\frac{268 + x^2}{8} - \left(\frac{42 + x}{8}\right)^2 = 2$$

$$8(268 + x^2) - (42 + x)^2 = 128$$

$$380 + 7x^2 - 84x = 128$$

$$7x^2 - 84x + 252 = 0$$

$$7(x^2 - 12x + 36) = 0$$

$$x^2 - 12x + 36 = 0$$

$$(x - 6)^2 = 0$$

$$x = 6$$

Question 15

a Using the calculator's statistical functions:

$m = -0.9696\ldots \approx -0.97$

$c = 80.9044\ldots \approx 80.90$

The least-squares regression line has equation $y = -0.97x + 80.90$.

b The equation comes from data of people aged 0–40; but Dom's father is 79, so this equation might not be valid for him.

c Dom: $y = 80.90 - 0.97 \times 36 = 45.98$

Approx. 46 years

Question 16

$6 \times 11 = 66$

Let x = Sarah's individual score

$$\frac{66 + x}{7} = 12$$

$$x = 18$$

Question 17

a IQR = $Q_3 - Q_1 = 155 - 40 = 115$

b Chatter: data is positively skewed; IQR = 115 (middle 50% of data 40–155), median = 80

InstaTik: data is symmetrical, IQR = $145 - 75 = 70$ (middle 50% of data 75–145), median = 110

Chatter has a positive skew with a high tail value of 250, median of 80, and a greater interquartile range of 115 (more spread) than InstaTik. InstaTik has symmetric data with a higher median (110) and smaller IQR of 70. InstaTik is used slightly more by this Year 12 group than Chatter.

Question 18

a $r = 0.9068\ldots \approx 0.907$

b Let x be the marks for Drama and y be the marks for Biology.

$m = 1.3758\ldots \approx 1.38$

$c = -29.4843\ldots \approx -29.48$

$y = 1.38x - 29.48$

Question 19

a As the number of energy drinks consumed increases, the number of hours sleeping decreases.

b Gradient $(m) = r \times \dfrac{\text{standard deviation of } y \text{ scores}}{\text{standard deviation of } x \text{ scores}}$

$$= -0.770 \times \frac{1.12}{1.56}$$

$$= -0.5528\ldots$$

$$\approx -0.55$$

The amount of sleep decreases by 0.55 h for each additional can of energy drink consumed.

Question 20

$\bar{x} = 22 - 0.525 = 21.475$

$\bar{y} = \dfrac{684}{20} = 34.2$

Substitute $\bar{x}$ and $\bar{y}$ into $y = -10.6063 + bx$

$34.2 = -10.6063 + b \times 21.475$

$b = \dfrac{44.8063}{21.475}$

$ = 2.086\,44\ldots$

$x = 19$, $y = -10.6063 + 2.086\,44\ldots \times 19$

$ = 29.036\,06\ldots$

In a 15-second interval, 29 chirps are expected.

WORKED SOLUTIONS

HSC exam topic grid (2011–2020)

This table shows the coverage of this topic in past HSC exams by question number. The past exams can be downloaded from the NESA website (www.educationstandards.nsw.edu.au) by selecting 'Year 11 – Year 12', 'HSC exam papers'. NESA marking feedback and guidelines can also be found there.

Statistics and bivariate data were introduced to the Mathematics Advanced course in 2020, but both are common content with Mathematics Standard 2. In the table below, * refers to past HSC questions (2011–2019) in Mathematics Standard 2 and Mathematics General 2. Before 2019, 'Mathematics Standard 2' was called 'Mathematics General 2', and before 2014, 'General Mathematics'. For these exams, select 'Year 11 – Year 12', 'Resources archive', 'HSC exam papers archive'.

	Displaying and analysing data	Comparing data	Scatterplots and correlation	Linear regression
2011*	7, 14, 17, 25(a)(i), 25(d)	11, 25(b)	8	
2012*	1, 26(e)	28(d)	11, 29(a)	19 (replace 'median regression' with 'line of best fit')
2013*	**6**, 14, 15, 26(b), 27(c), 29(b)(i)	**26(f)**	2	28(b)
2014*	14, 26(e), 30(b)(ii)–(iv)	29(c)	30(b)(i)	30(b)(vi)–(viii)
2015*	4, 6, 27(d), 29(d)		28(e)	28(e)
2016*	7, 19, 21, 27(c)	22, 29(c)	3, 29(d)(i)	29(d)(ii) (use calculator and table of values)
2017*	1, 27(a), 29(d), 30(a)		12	
2018*	1, 3, 6, 11, 26(e)	26(d)		29(d) (use calculator and table of values for (i))
2019*	10, 19, 39	39	23(a)–(b)	23(c)
2020 new course	**27**			**27**

Questions in **bold** can be found in this chapter.

CHAPTER 7
PROBABILITY DISTRIBUTIONS

PROBABILITY DISTRIBUTIONS

Continuous probability distributions

Probability density function (PDF)

$$\int_{-\infty}^{\infty} f(x)\,dx = 1$$

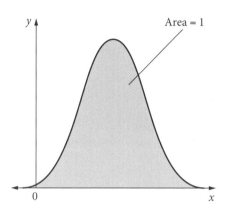

Area = 1

$$P(a \le X \le b) = \int_{a}^{b} f(x)\,dx$$

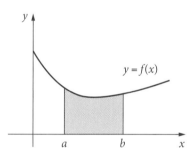

$$y = f(x)$$

Cumulative distribution function (CDF)

$$F(x) = \int_{a}^{x} f(x)\,dx$$

Quantiles

- Median: $\int_{a}^{x} f(x)\,dx = \frac{1}{2}$

- Lower quartile: $\int_{a}^{x} f(x)\,dx = \frac{1}{4}$

- Upper quartile: $\int_{a}^{x} f(x)\,dx = \frac{3}{4}$

- Deciles: 3rd decile > bottom 30% of values
  $\rightarrow \int_{a}^{x} f(x)\,dx = 0.3$

- Percentiles: 64th percentile > bottom 64% of values
  $\rightarrow \int_{a}^{x} f(x)\,dx = 0.64$

Normal distribution 📎

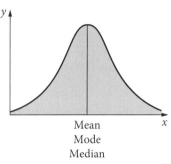

Mean
Mode
Median

z-scores 📎

- Measures number of standard deviations from the mean: $z = -1.6$ means 1.6 standard deviations below the mean.

$$z = \frac{x - \mu}{\sigma}$$

- **Empirical rule**

Standard deviations

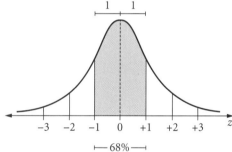

68%

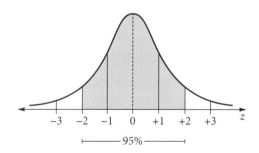

95%

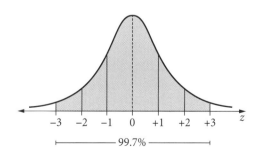

99.7%

- Probability tables for z-scores
- Comparing z-scores

Glossary

continuous probability distribution
A distribution of a continuous random variable. It is represented by a **probability density function** $P(X = x)$ or $p(x)$, where X is the random variable.

continuous random variable
A random variable that can take any value; for example, the amount of rainfall or the height of a basketball player.

cumulative distribution function (CDF)
Given a continuous random variable X, the function $F(x)$ for the probability $P(X \leq x)$.

empirical rule
A statistical rule which states that for a **normal distribution**, almost all observed data will fall within three standard deviations (σ) of the mean or average (μ).

In particular, the empirical rule predicts that approximately:

- 68% of data falls within the first standard deviation ($\mu \pm \sigma$)

- 95% within the first two standard deviations ($\mu \pm 2\sigma$)

- 99.7% within the first three standard deviations ($\mu \pm 3\sigma$).

normal distribution
A bell-shaped symmetrical continuous distribution.

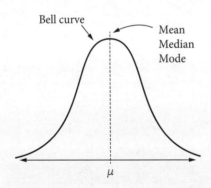

The mean, median and mode are equal and at the centre of the distribution.

probability density function (PDF)
A function $P(X = x)$ or $p(x)$ of a continuous random variable, whose integral across an interval gives the probability that the value of the variable lies within the same interval. For a continuous probability function, the probability of a single outcome cannot be found. Only probabilities for a *range* of values can be calculated.

random variable
A variable whose values are based on a chance experiment.

uniform probability distribution
A probability distribution in which every outcome has the same probability.

z-score
A statistical value of how many standard deviations a data value is above or below the mean. Values above the mean have positive z-scores. Values below the mean have negative z-scores. The mean has a z-score of 0.

9780170459228

Topic summary

Random variables (MA-S3)

S3.1 Continuous random variables

Probability density functions

A **probability density function (PDF)** gives the probability that a random variable will have a value within a specific interval.

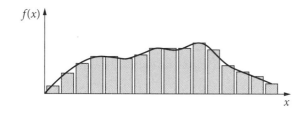

The area under a PDF is 1.

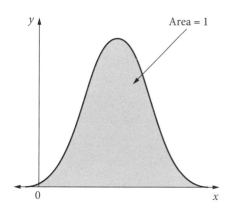

$$\int_{-\infty}^{\infty} f(x)\,dx = 1,$$

where $f(x) \geq 0$ (since $0 \leq p(x) \leq 1$).

A **continuous probability distribution** is represented by a function $P(X = x)$ or $p(x)$, called a PDF, where X is the random variable. For a continuous probability function, the probability of a single outcome cannot be found; that is $P(X = x) = 0$. Only probabilities for a *range* of values can be calculated.

Probabilities in a PDF

$P(X \leq x) = \int_{a}^{x} f(x)\,dx,$

where $y = f(x)$ is a PDF defined for $[a, b]$.

$P(a \leq X \leq b) = \int_{a}^{b} f(x)\,dx,$

where $y = f(x)$ is a PDF defined for $[a, b]$.

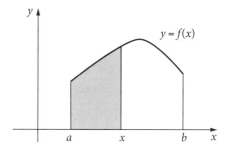

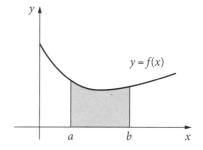

Example 1   ©NESA   2020 HSC EXAM, QUESTION 23

A continuous random variable, X, has the following probability density function.

$$f(x) = \begin{cases} \sin x & \text{for } 0 \leq x \leq k \\ 0 & \text{for all other values of } x \end{cases}$$

a Find the value of k.

b Find $P(X \leq 1)$, correct to four decimal places.

Solution

a
$$\int_0^k \sin x \, dx = 1$$
$$\left[-\cos x\right]_0^k = 1$$
$$-\cos k - (-\cos 0) = 1$$
$$-\cos k + 1 = 1$$
$$-\cos k = 0$$
$$\cos k = 0$$
$$k = \frac{\pi}{2}$$

b $P(X \leq 1) = \int_0^1 \sin x \, dx$
$$= -\cos 1 + 1 \quad \text{from part } \mathbf{a}$$
$$= 0.459\,697\ldots$$
$$\approx 0.4597$$

In a **uniform probability distribution**, every outcome has the same probability.

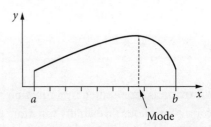

$$f(x) = \begin{cases} \dfrac{1}{b-a} & \text{for } a \leq x \leq b \\ 0 & \text{for all other } x \text{ values} \end{cases}$$

The **mode** is the highest point of the probability density function (PDF).

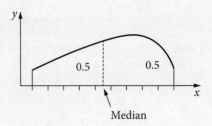

The **median** lies at the point such that $\int_a^x f(x)\,dx = \dfrac{1}{2}$, where $y = f(x)$ is a PDF defined for $[a, b]$.

Quantiles

Quartiles

$$\text{Lower quartile: } \int_a^x f(x)\,dx = \frac{1}{4}$$

$Q_1 >$ bottom 25% of scores

$$\text{Upper quartile: } \int_a^x f(x)\,dx = \frac{3}{4}$$

$Q_3 >$ bottom 75% of scores

Deciles and percentiles

Deciles divide data sets into 10 equal parts. For example, the 3rd decile > bottom 30% of scores.
Percentiles divide data sets into 100 equal parts. For example, the 64th percentile > bottom 64% of scores.

$$\int_a^x f(x)\,dx = 0.64$$

Cumulative distribution function

For a **cumulative distribution function** (CDF), the probability for a range of values can be calculated by

$$P(X \le x) = F(x) = \int_a^b f(x)\,dx, \text{ where } y = f(x) \text{ is a PDF defined for } [a, b].$$

Example 2 ©NESA SAMPLE 2020 HSC EXAM, QUESTION 31

A bid made at an auction for a real estate property, in millions of dollars, can be modelled by the random variable X with the probability density function

$$f(x) = \begin{cases} k(16 - x^2) & \text{for } 1 \le x \le 4 \\ 0 & \text{otherwise} \end{cases}$$

a Show that the value of k is $\dfrac{1}{27}$.

b Find the cumulative distribution function.

c Find the probability that a bid of more than 3 million dollars will be made.

Solution

a
$$\int_1^4 k(16 - x^2)\,dx = 1$$

$$k\left[16x - \frac{x^3}{3}\right]_1^4 = 1$$

$$k\left[\left(16(4) - \frac{4^3}{3}\right) - \left(16 - \frac{1}{3}\right)\right] = 1$$

$$k\left(\frac{128}{3} - \frac{47}{3}\right) = 1$$

$$27k = 1$$

$$k = \frac{1}{27}$$

b
$$F(x) = \int_1^x \frac{1}{27}(16 - x^2)\,dx$$

$$= \frac{1}{27}\left[16x - \frac{x^3}{3}\right]_1^x$$

$$= \frac{1}{27}\left[\left(16x - \frac{x^3}{3}\right) - \left(16 - \frac{1}{3}\right)\right]$$

$$= \frac{1}{27}\left(16x - \frac{x^3}{3} - \frac{47}{3}\right)$$

$$= \frac{1}{81}(-x^3 + 48x - 47) \qquad \text{for } 1 \le x \le 4$$

c
$$P(X > 3) = 1 - P(X < 3)$$

$$= 1 - \frac{1}{81}[-3^3 + 48(3) - 47]$$

$$= 1 - \frac{1}{81}(70)$$

$$= \frac{11}{81}$$

TOPIC SUMMARY

Common content with Mathematics Standard 2 course

S3.2 The normal distribution

A normal distribution is a special probability distribution that is symmetrical in shape. It is commonly known as a bell curve, due to its shape.

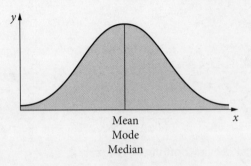

Mean
Mode
Median

Mean = median = mode

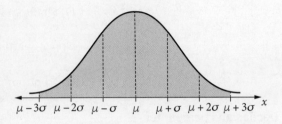

Most data (about 99.7%) lies within 3 standard deviations, σ, of the mean, μ.

z-scores

A **z-score** (or standardised score) shows the position of a 'raw' score relative to the mean.

It is the number of standard deviations a data value is from the mean.

For example, a z-score of 1.8 means a score is 1.8 standard deviations above the mean. Positive z-scores are *above* the mean.

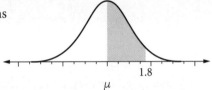

A z-score of −0.5 means a score is half a standard deviation below the mean. Negative z-scores are *below* the mean.

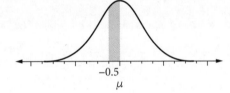

A z-score of 0 means the score is the mean.

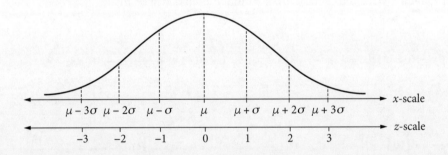

$$z\text{-score} = \frac{\text{score} - \text{mean}}{\text{standard deviation}}$$

$$z = \frac{x - \mu}{\sigma}$$

The empirical rule

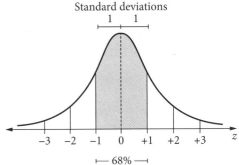

Standard deviations

> **Hint**
> The z-score formula and the empirical rule are printed on the HSC exam formula sheet and at the back of this book.

Approximately 68% of values are within 1 standard deviation of the mean and $z = -1$ to $z = 1$.

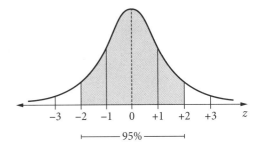

Approximately 95% of values are within 2 standard deviations of the mean and $z = -2$ to $z = 2$.

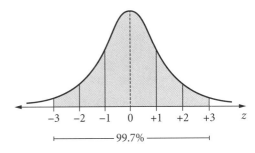

Approximately 99.7% of values are within 3 standard deviations of the mean and $z = -3$ to $z = 3$.

> **Hint**
> Instead of using the empirical rule, some students like to memorise the areas of the individual regions to make calculations easier in problems. The area from 0 to 1 is 34% ($\frac{1}{2}$ of 68%), then moving right it's 13.5%, 2.35% and 0.15%.
> It is the same going from 0 to −1 and so on.

TOPIC SUMMARY

Example 3 ©NESA 2020 HSC EXAM, QUESTION 28(a)

In a particular country, the hourly rate of pay for adults who work is normally distributed with a mean of $25 and a standard deviation of $5.

Two adults who both work are chosen at random.

Find the probability that at least one of them earns between $15 and $30 per hour.

Solution

Normal distribution with mean $25 and standard deviation $5, so curve looks as shown on the right.

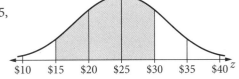

The probability of selecting an adult who earns between $15 and $30 per hour is shaded.

By the empirical rule, the $20–$30 region is 68%.

The $15–$20 region is (95% − 68%) ÷ 2 = 13.5%.

So P(adult earns $15–$30) = 68% + 13.5% = 81.5% = 0.815

and P(adult doesn't earn $15–$30) = 1 − 0.815 = 0.185.

P(at least 1 adult earns $15–$30) = 1 − P(no adult earns $15–$30)

$$= 1 - 0.185^2$$
$$= 0.965\,775$$

Probability table for z-scores

For decimal values of z, we can look up areas and percentages in a table. In this table, values represent the area to the left of (or less than) the z-score.

Row labels show the z-score to one decimal place. Column labels show the second decimal place.

For example, $z = -2.85 = -2.8 - 0.05$ has a value of 0.0022, so 0.0022 or 0.22% of scores are below $z = -2.85$.

z	0.00	0.01	0.02	0.03	0.04	0.05	0.06	0.07	0.08	0.09
−3.4	0.0003	0.0003	0.0003	0.0003	0.0003	0.0003	0.0003	0.0003	0.0003	0.0002
−3.3	0.0005	0.0005	0.0005	0.0004	0.0004	0.0004	0.0004	0.0004	0.0004	0.0003
−3.2	0.0007	0.0007	0.0006	0.0006	0.0006	0.0006	0.0006	0.0005	0.0005	0.0005
−3.1	0.0010	0.0009	0.0009	0.0009	0.0008	0.0008	0.0008	0.0008	0.0007	0.0007
−3.0	0.0013	0.0013	0.0013	0.0012	0.0012	0.0011	0.0011	0.0011	0.0010	0.0010
−2.9	0.0019	0.0018	0.0018	0.0017	0.0016	0.0016	0.0015	0.0015	0.0014	0.0014
−2.8	0.0026	0.0025	0.0024	0.0023	0.0023	0.0022	0.0021	0.0021	0.0020	0.0019
−2.7	0.0035	0.0034	0.0033	0.0032	0.0031	0.0030	0.0029	0.0028	0.0027	0.0026
−2.6	0.0047	0.0045	0.0044	0.0043	0.0041	0.0040	0.0039	0.0038	0.0037	0.0036
−2.5	0.0062	0.0060	0.0059	0.0057	0.0055	0.0054	0.0052	0.0051	0.0049	0.0048
−2.4	0.0082	0.0080	0.0078	0.0075	0.0073	0.0071	0.0069	0.0068	0.0066	0.0064
−2.3	0.0107	0.0104	0.0102	0.0099	0.0096	0.0094	0.0091	0.0089	0.0087	0.0084
−2.2	0.0139	0.0136	0.0132	0.0129	0.0125	0.0122	0.0119	0.0116	0.0113	0.0110
−2.1	0.0179	0.0174	0.0170	0.0166	0.0162	0.0158	0.0154	0.0150	0.0146	0.0143
−2.0	0.0228	0.0222	0.0217	0.0212	0.0207	0.0202	0.0197	0.0192	0.0188	0.0183
−1.9	0.0287	0.0281	0.0274	0.0268	0.0262	0.0256	0.0250	0.0244	0.0239	0.0233
−1.8	0.0359	0.0351	0.0344	0.0336	0.0329	0.0322	0.0314	0.0307	0.0301	0.0294
−1.7	0.0446	0.0436	0.0427	0.0418	0.0409	0.0401	0.0392	0.0384	0.0375	0.0367
−1.6	0.0548	0.0537	0.0526	0.0516	0.0505	0.0495	0.0485	0.0475	0.0465	0.0455
−1.5	0.0668	0.0655	0.0643	0.0630	0.0618	0.0606	0.0594	0.0582	0.0571	0.0559
−1.4	0.0808	0.0793	0.0778	0.0764	0.0749	0.0735	0.0721	0.0708	0.0694	0.0681
−1.3	0.0968	0.0951	0.0934	0.0918	0.0901	0.0885	0.0869	0.0853	0.0838	0.0823
−1.2	0.1151	0.1131	0.1112	0.1093	0.1075	0.1056	0.1038	0.1020	0.1003	0.0985
−1.1	0.1357	0.1335	0.1314	0.1292	0.1271	0.1251	0.1230	0.1210	0.1190	0.1170
−1.0	0.1587	0.1562	0.1539	0.1515	0.1492	0.1469	0.1446	0.1423	0.1401	0.1379
−0.9	0.1841	0.1814	0.1788	0.1762	0.1736	0.1711	0.1685	0.1660	0.1635	0.1611
−0.8	0.2119	0.2090	0.2061	0.2033	0.2005	0.1977	0.1949	0.1922	0.1894	0.1867
−0.7	0.2420	0.2389	0.2358	0.2327	0.2296	0.2266	0.2236	0.2206	0.2177	0.2148
−0.6	0.2743	0.2709	0.2676	0.2643	0.2611	0.2578	0.2546	0.2514	0.2483	0.2451
−0.5	0.3085	0.3050	0.3015	0.2981	0.2946	0.2912	0.2877	0.2843	0.2810	0.2776
−0.4	0.3446	0.3409	0.3372	0.3336	0.3300	0.3264	0.3228	0.3192	0.3156	0.3121
−0.3	0.3821	0.3783	0.3745	0.3707	0.3669	0.3632	0.3594	0.3557	0.3520	0.3483
−0.2	0.4207	0.4168	0.4129	0.4090	0.4052	0.4013	0.3974	0.3936	0.3897	0.3859
−0.1	0.4602	0.4562	0.4522	0.4483	0.4443	0.4404	0.4364	0.4325	0.4286	0.4247
−0.0	0.5000	0.4960	0.4920	0.4880	0.4840	0.4801	0.4761	0.4721	0.4681	0.4641

z	0.00	0.01	0.02	0.03	0.04	0.05	0.06	0.07	0.08	0.09
0.0	0.5000	0.5040	0.5080	0.5120	0.5160	0.5199	0.5239	0.5279	0.5319	0.5359
0.1	0.5398	0.5438	0.5478	0.5517	0.5557	0.5596	0.5636	0.5675	0.5714	0.5753
0.2	0.5793	0.5832	0.5871	0.5910	0.5948	0.5987	0.6026	0.6064	0.6103	0.6141
0.3	0.6179	0.6217	0.6255	0.6293	0.6331	0.6368	0.6406	0.6443	0.6480	0.6517
0.4	0.6554	0.6591	0.6628	0.6664	0.6700	0.6736	0.6772	0.6808	0.6844	0.6879
0.5	0.6915	0.6950	0.6985	0.7019	0.7054	0.7088	0.7123	0.7157	0.7190	0.7224
0.6	0.7257	0.7291	0.7324	0.7357	0.7389	0.7422	0.7454	0.7486	0.7517	0.7549
0.7	0.7580	0.7611	0.7642	0.7673	0.7704	0.7734	0.7764	0.7794	0.7823	0.7852
0.8	0.7881	0.7910	0.7939	0.7967	0.7995	0.8023	0.8051	0.8078	0.8106	0.8133
0.9	0.8159	0.8186	0.8212	0.8238	0.8264	0.8289	0.8315	0.8340	0.8365	0.8389
1.0	0.8413	0.8438	0.8461	0.8485	0.8508	0.8531	0.8554	0.8577	0.8599	0.8621
1.1	0.8643	0.8665	0.8686	0.8708	0.8729	0.8749	0.8770	0.8790	0.8810	0.8830
1.2	0.8849	0.8869	0.8888	0.8907	0.8925	0.8944	0.8962	0.8980	0.8997	0.9015
1.3	0.9032	0.9049	0.9066	0.9082	0.9099	0.9115	0.9131	0.9147	0.9162	0.9177
1.4	0.9192	0.9207	0.9222	0.9236	0.9251	0.9265	0.9279	0.9292	0.9306	0.9319
1.5	0.9332	0.9345	0.9357	0.9370	0.9382	0.9394	0.9406	0.9418	0.9429	0.9441
1.6	0.9452	0.9463	0.9474	0.9484	0.9495	0.9505	0.9515	0.9525	0.9535	0.9545
1.7	0.9554	0.9564	0.9573	0.9582	0.9591	0.9599	0.9608	0.9616	0.9625	0.9633
1.8	0.9641	0.9649	0.9656	0.9664	0.9671	0.9678	0.9686	0.9693	0.9699	0.9706
1.9	0.9713	0.9719	0.9726	0.9732	0.9738	0.9744	0.9750	0.9756	0.9761	0.9767
2.0	0.9772	0.9778	0.9783	0.9788	0.9793	0.9798	0.9803	0.9808	0.9812	0.9817
2.1	0.9821	0.9826	0.9830	0.9834	0.9838	0.9842	0.9846	0.9850	0.9854	0.9857
2.2	0.9861	0.9864	0.9868	0.9871	0.9875	0.9878	0.9881	0.9884	0.9887	0.9890
2.3	0.9893	0.9896	0.9898	0.9901	0.9904	0.9906	0.9909	0.9911	0.9913	0.9916
2.4	0.9918	0.9920	0.9922	0.9925	0.9927	0.9929	0.9931	0.9932	0.9934	0.9936
2.5	0.9938	0.9940	0.9941	0.9943	0.9945	0.9946	0.9948	0.9949	0.9951	0.9952
2.6	0.9953	0.9955	0.9956	0.9957	0.9959	0.9960	0.9961	0.9962	0.9963	0.9964
2.7	0.9965	0.9966	0.9967	0.9968	0.9969	0.9970	0.9971	0.9972	0.9973	0.9974
2.8	0.9974	0.9975	0.9976	0.9977	0.9977	0.9978	0.9979	0.9979	0.9980	0.9981
2.9	0.9981	0.9982	0.9982	0.9983	0.9984	0.9984	0.9985	0.9985	0.9986	0.9986
3.0	0.9987	0.9987	0.9987	0.9988	0.9988	0.9989	0.9989	0.9989	0.9990	0.9990
3.1	0.9990	0.9991	0.9991	0.9991	0.9992	0.9992	0.9992	0.9992	0.9993	0.9993
3.2	0.9993	0.9993	0.9994	0.9994	0.9994	0.9994	0.9994	0.9995	0.9995	0.9995
3.3	0.9995	0.9995	0.9995	0.9996	0.9996	0.9996	0.9996	0.9996	0.9996	0.9997
3.4	0.9997	0.9997	0.9997	0.9997	0.9997	0.9997	0.9997	0.9997	0.9997	0.9998

TOPIC SUMMARY

Comparing z-scores

Example 4   ©NESA 2015 MATHEMATICS GENERAL 2 HSC EXAM, QUESTION 28(b)

The results of two tests are normally distributed. The mean and standard deviation for each test are displayed in the table.

	Mathematics	English
$\bar{x}$	70	75
s	6.5	8

Kristoff scored 74 in Mathematics and 80 in English. He claims that he has performed better in English.

Is Kristoff correct? Justify your answer using appropriate calculations.

Solution

Convert both results to z-scores to compare.

Mathematics $z = \dfrac{74 - 70}{6.5} = 0.615\ldots$

English $z = \dfrac{80 - 75}{8} = 0.625$

Kristoff did better in English because its z-score is higher.

Practice set 1

Multiple-choice questions

Solutions start on page 187.

Question 1 ⬤◐◐

Which one of the following histograms shows data that could represent a normal distribution?

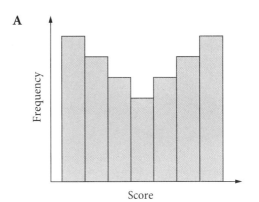

A

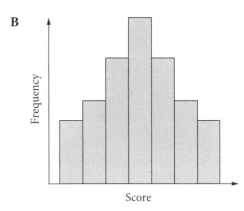

B

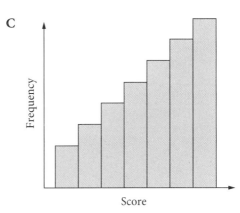

C

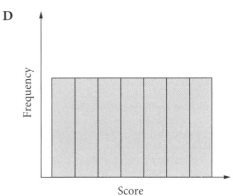

D

Question 2 ⬤◐◐

A machine produces metal rods. The mean of the lengths of the rods is 45 cm and the standard deviation is 0.05 cm.

Assuming a normal distribution, what percentage of metal rods will have a length of between 44.95 cm and 45.05 cm?

A 34%

B 68%

C 95%

D 99.7%

Question 3 ©NESA 2020 HSC EXAM, QUESTION 3   ○●●

John recently did a class test in each of three subjects. The class scores on each test were normally distributed.

The table shows the subjects and John's scores as well as the mean and standard deviation of the class scores on each test.

Subject	John's score	Mean	Standard deviation
French	82	70	8
Commerce	80	65	5
Music	74	50	12

Relative to the rest of the class, which row of the table below shows John's strongest subject and his weakest subject?

	Strongest subject	Weakest subject
A	Commerce	French
B	French	Music
C	Music	French
D	Commerce	Music

Question 4 ○●●

Mitchell and Zoe both play in the cricket competition. When the heights of all players in the competition are considered, Mitchell has a height of $z = 0.98$ and Zoe has a height of $z = 1.36$.

Which one of the following statements is NOT true?

A Mitchell and Zoe are both above the mean height for cricket players in their competition.

B Zoe is taller than Mitchell.

C Mitchell is 0.38 metres shorter than Zoe.

D More than 50% of the players are taller than Mitchell.

Question 5 ○●●

The graph of the probability density function of a continuous random variable, X, is shown below.

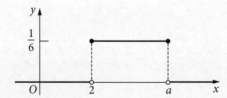

Find the value of a.

A 3

B 4

C 6

D 8

Question 6

Each of the diagrams shown below represents the normal distribution with mean $\mu = 0$ and standard deviation $\sigma = 1$.

Z is a normally distributed random variable with mean $\mu = 0$ and standard deviation $\sigma = 1$.

Which graph best demonstrates $P(Z \leq 1)$?

A

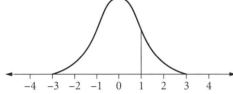

B

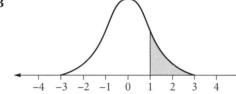

C

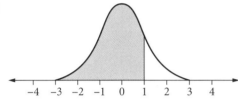

D

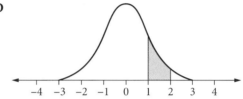

Question 7

A continuous probability distribution $y = f(x)$ in the domain $[1, 10]$ is shown below.

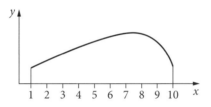

Which one of the following statements is correct?

A The median is 7.7.　　**B** The mode is 7.7.　　**C** The median is 6.5.　　**D** The mode is 6.5.

The information given below is used in Questions 8, 9 and 10.

The pulse rates of a population of Year 12 students are approximately normally distributed with a mean of 71 beats per minute and a standard deviation of 4 beats per minute.

Question 8

Georgia is selected randomly from this population. She has a z-score of $z = -2.5$. What is her pulse rate?

A 61　　　　　　**B** 67　　　　　　**C** 75　　　　　　**D** 81

Question 9

Zara is another student randomly selected from this population. Zara has a z-score of $z = 1$. What is the percentage of students with a pulse rate that is higher than Zara's?

A 16%　　　　　**B** 24%　　　　　**C** 34%　　　　　**D** 68%

Question 10

What percentage of students from this Year 12 population have a pulse rate of less than 63 beats per minute or greater than 75 beats per minute?

A 18.5%　　　　**B** 36.5%　　　　**C** 50%　　　　　**D** 95%

Question 11 ●●

A function is given by $f(x) = ax^2$, defined for the domain $[0, 6]$. What is the value of a for which a probability density function exists?

A $\dfrac{1}{648}$ B $\dfrac{1}{432}$ C $\dfrac{1}{72}$ D $\dfrac{1}{36}$

Question 12 ●●

Use the probability table for z-scores on pages 174–175 to find the value of m such that $P(1.14 \leq Z \leq 1.40) = m$, given that Z is a standard normal random variable.

A 0.0157 B 0.0463 C 0.1429 D 0.9843

Question 13 ●●

A set of scores is normally distributed. A score of 23 has a z-score of $z = -1$ and a score of 41 has a z-score of $z = 3$. What is the mean of this set of scores?

A 16 B 24.5 C 27.5 D 32

Question 14 ●●

The probability density function of the continuous random variable, X, is given by

$$f(x) = \begin{cases} p & -3 \leq x \leq 0 \\ 3p & 1 \leq x \leq 12p \\ 0 & \text{otherwise} \end{cases}$$

What is the value of p?

A $\dfrac{1}{6}$ B $\dfrac{1}{4}$ C $\dfrac{1}{3}$ D $\dfrac{1}{2}$

Question 15 ●●

The lifetime, X, in tens of thousands of flying hours, of wings on a certain aeroplane is modelled by the probability density function

$$f(x) = \begin{cases} \dfrac{1}{9}x(5 - x) & 1 \leq x \leq 6 \\ 0 & \text{otherwise} \end{cases}$$

What is $P(X > 3.5)$?

A $\dfrac{1}{27}$ B $\dfrac{5}{27}$ C $\dfrac{4}{9}$ D $\dfrac{5}{9}$

Question 16 ●●

For a normally distributed set of scores, which one of the following statements is *false*?

A The z-score describes how far an individual score is from the mean.

B The distribution is symmetrical about the mean.

C The mean and median of the distribution are equal.

D A z-score greater than 3 has a higher probability than a z-score greater than 2.

PRACTICE SET 1

Question 17 ⬤⬤◯

The heights of an adult female population are normally distributed with a mean of 1.62 m and a standard deviation of 9.5 cm.

Sarah is in Year 8 and goes to the doctor for a check-up. She is told she is in the 80th percentile for height. Assuming that Sarah remains at the 80th percentile, which of these heights would her estimated height as an adult be closest to?

Use the probability table on pages 174–175 to answer this question.

A 1.65 m   **B** 1.69 m   **C** 1.70 m   **D** 1.76 m

Question 18 ⬤⬤◯

Let X be a continuous random variable with probability density function

$$f(x) = \begin{cases} \dfrac{ax}{54} & 26 \leq x \leq 28 \\ 0 & \text{otherwise} \end{cases}$$

Evaluate $P(26.5 \leq x \leq 27.5)$.

A $\dfrac{2}{5}$ **B** $\dfrac{25}{54}$ **C** $\dfrac{1}{2}$ **D** $\dfrac{5}{9}$

Question 19 ©NESA 2020 HSC EXAM, QUESTION 9 ⬤⬤⬤

Suppose the weight of melons is normally distributed with a mean of μ and a standard deviation of σ.

A melon has a weight below the lower quartile of the distribution but NOT in the bottom 10% of the distribution.

Which of the following most accurately represents the region in which the weight of this melon lies?

A
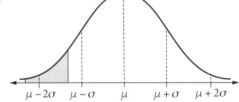

B
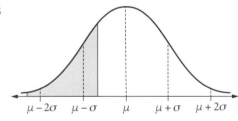

C
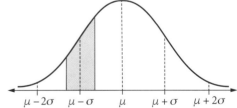

D
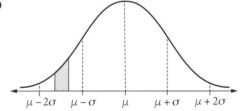

Question 20 ⬤⬤⬤

A probability density function $y = f(x)$ is given by

$$f(x) = \begin{cases} \cos x + 1 & k < x < k+1 \\ 0 & \text{elsewhere} \end{cases}$$

where $0 < k < 2$.

What is the value of k?

A 1   **B** $\dfrac{\pi-1}{2}$ **C** $\pi - 1$ **D** $\dfrac{\pi}{2}$

Practice set 2

Short-answer questions

Solutions start on page 189.

Question 1 (5 marks) ⬤⬤⬜

Zac is a Year 12 student. He records the time it takes him to travel home from school each day.

The histogram of relative frequencies shows the times he recorded.

Use the data above to estimate each answer.

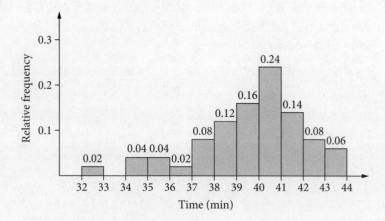

a What is the probability that Zac's next trip home from school takes him:

   i less than 36 minutes?   1 mark

   ii at least 35 minutes but at most 39 minutes?   2 marks

b On 3 consecutive days, Zac needs to be at school no later than 8:40 am. If he leaves home at 8:02 am each day, find the probability he will be on time or early on all 3 days.   2 marks

Question 2 (1 mark) ⬤⬜⬜

The heights of Year 12 girls are normally distributed with a mean of 166 cm and standard deviation of 6 cm.

Calculate the z-score for a height of 151 cm.   1 mark

Question 3 (1 mark) ⬤⬜⬜

State the mode of the continuous probability distribution.   1 mark

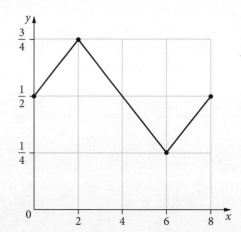

Question 4 (2 marks) ⬤⬤⬜

A standard normal distribution has a mean of 70 and a standard deviation of 8.

Find the probability that a value taken at random from the population lies between 70 and 74.   2 marks

Question 5 (2 marks) ⬤⬜⬜

A company produces 1 kg packets of flour. A quality control check found that the weights of the packets were normally distributed with a mean weight of 1005 g and a standard deviation of 3 g. The company rejects packets of flour that weigh more than 2 standard deviations from the mean.

a What percentage of packets of flour will be rejected for sale?   1 mark

b From 1200 packets of flour produced, how many packets are expected to be rejected?   1 mark

Question 6 (6 marks) ▣▣▢

Phoebe plays indoor hockey as goalkeeper and is practising with the help of a machine. The machine randomly shoots a hockey ball along the ground at or near the goal, which is 3 metres wide. The machine has an equal chance of shooting the ball so that the centre of the hockey ball crosses the goal line anywhere between point A, which is 5 metres left of the goal, and point B, which is 7 metres right of the goal, as shown in the diagram below.

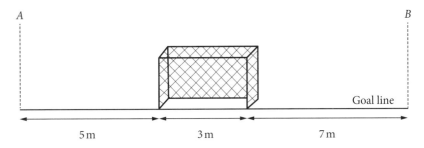

Phoebe does a practice run with no one standing in the goal area. Do not consider the width of the goal posts in this question.

Let the random variable X represent the distance the centre of the ball crosses the goal line to the right of point A.

a Graph the probability density function for the random variable X.   2 marks

b Calculate the probability that the machine shoots a ball so that its centre misses the goal   1 mark
    to the right.

c What is the probability that the machine shoots a ball so that its centre is inside the goal?   1 mark

d If the machine shoots a ball so that its centre misses the goal, what is the probability that   2 marks
    the ball's centre misses to the left?

Question 7 (6 marks) ▣▣▢

A machine in a soft drink factory fills bottles and cans. The machine is set to fill 375 mL cans with a mean of 380 mL. The contents of the cans are normally distributed with a standard deviation of 2 mL.

a What percentage of the cans would you expect to contain:

  i between 378 mL and 382 mL?   1 mark

  ii less than 376 mL?   1 mark

  iii less than 376 mL and greater than 382 mL?   2 marks

b The manufacturers at this factory state that most of their soft drink cans contain at least
    375 mL, as shown on the cans.

    Explain why this statement is correct. Show working to justify your answer.   2 marks

Question 8 (4 marks) ▣▢▢

Use the probability table on pages 174–175 to answer this question.

Nelson Secondary College opens at 7:30 am on each school day. The daily arrival times of students at the school follow a normal distribution. The mean arrival time is 35 minutes after the school opens and the standard deviation is 5 minutes.

A student is chosen at random. Calculate the probability that the student arrives:

a at least 30 minutes after school opens.   2 marks

b between 28 minutes and 38 minutes after the school opens.   2 marks

Question 9 (2 marks) ●●▮

The results of Ashna's Science and Geography tests are normally distributed. The mean and standard deviation for each of the tests are shown in the table.

	Science	Geography
$\bar{x}$	75	69
σ	8	6.5

Ashna scored 81 in Science and 74 in Geography. She tells her friend Bella, 'Of these 2 results, my mark was better in Science'.

Is Ashna correct? Show all possible working to justify your answer.   2 marks

Question 10 (4 marks) ●▮▮

An IT company is required to digitise an audio signal. The error, X, in digitising an audio signal, has a uniform distribution with the probability density function given by

$$f(x) = \begin{cases} 1 & -0.5 < x < 0.5 \\ 0 & \text{otherwise} \end{cases}$$

a Sketch the graph of $y = f(x)$ for the given domain.   1 mark

b What is the probability that the error is at least 0.45 (that is, $P(X > 0.45)$)?   1 mark

c Given that the error is negative, what is the probability that the error is less than −0.45   2 marks
(that is, calculate $P(X < -0.45 \mid X < 0)$)?

Question 11 (6 marks) ●●▮

Use the probability table for standard normal distribution on pages 174–175 to answer this question.

The weight, in grams, of canned fruit is normally distributed with mean μ and standard deviation $\sigma = 7.8$.

a It is known that 10% of tinned fruit cans contain less than 200 g.

  i   Calculate the mean, μ.   2 marks

  ii What percentage of cans contain more than 225 g of fruit?   2 marks

b The machine settings are changed so that the weight, in grams, of canned fruit is normally   2 marks
distributed with mean $\mu = 205$ g and standard deviation σ. Given that 98% of cans contain
between 200 and 210 g of fruit, find the value of σ. Round your answer to two decimal places.

Question 12 (7 marks) ●●●

The function below is a probability density function for the given interval:

$$f(x) = \begin{cases} ax^2(x-2) & \text{for } 0 \le x \le 2 \\ 0 & \text{otherwise} \end{cases}$$

a Find the value of a.   3 marks

b Find the probability that $x \ge 1.2$.   2 marks

c Verify, using integration, that the median of this distribution exists for $x = 1.2285$.   2 marks

Question 13 (3 marks) ●●▮

The heights of a year group of 215 students are normally distributed with a mean of 166 cm and a standard deviation of 3 cm.

a What percentage of students are taller than 172 cm?   1 mark

b How many students are between 163 cm and 172 cm? 2 marks

Question 14 (3 marks) ⬤⬤▢

Waiting times for people at a virus testing clinic can be up to 4 hours. The probability density function is given below, where t is the waiting time in hours.

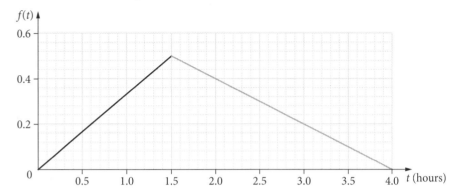

The graph above shows that

$$f(t) = \tfrac{1}{3}t, \, 0 \le t \le 1.5$$
$$\text{and } f(t) = -0.2t + 0.8, \, 1.5 \le t \le 4$$

a Calculate the probability that a person will wait for less than 1 hour to be tested. 1 mark

b Determine the probability that a person will wait between 1 hour and 3 hours to be tested. 2 marks

Question 15 (7 marks) ⬤⬤▢

The length of time, t minutes, spent waiting for an app to download has the following probability density function:

$$f(t) = \begin{cases} \tfrac{1}{2}(t-1) & 1 < t \le 2 \\ \tfrac{1}{16}(14t - 3t^2 - 8) & 2 < t \le 4 \\ 0 & \text{otherwise} \end{cases}$$

a Find the cumulative distribution function $F(t)$. 3 marks

b Find the 20th percentile of the time spent waiting for an app to download. 2 marks

c Calculate the probability that the time spent waiting for an app to download is more than 1.5 minutes. 2 marks

Question 16 (4 marks) ⬤⬤▢

A continuous random variable, X, has the following probability density function:

$$f(x) = \begin{cases} \cos x & \text{for } 0 \le x \le k \\ 0 & \text{for all other values} \end{cases}$$

a Find the value of k. 2 marks

b Find $P(X \le 1)$. Give your answer correct to four decimal places. 2 marks

Question 17 (4 marks) ©NESA 2019 SAMPLE HSC, QUESTION 13 ⬤⬤⬤

A continuous random variable X has a probability density function f given by

$$f(x) = \begin{cases} Ax + B & 1 \le x \le 4 \\ 0 & \text{elsewhere} \end{cases}$$

The median of X is 2. Find the values of A and B. 4 marks

Question 18 (5 marks) ●●○

Use the probability table on pages 174–175 for standard normal distribution to answer this question.

FastFoodz is a takeaway food company that runs a marketing campaign based on the delivery time of its products. FastFoodz claims that:

> **We deliver within a 10 km radius within 30 minutes of ordering**
>
> **Or**
>
> **YOUR ORDER IS**
> **FREE!**

The manager estimates that the actual time T, in minutes, from placing the order to delivery, is normally distributed with a mean of 25 minutes and a standard deviation of 4 minutes.

a Calculate the probability that takeaway food will be delivered for free.   1 mark

b FastFoodz is looking for ways to reduce the percentage of food delivered for free to 0.1%.   2 marks
If the manager thinks this can be achieved by increasing the advertised delivery time, what should the advertised delivery time be increased to?

c Additional training was provided by FastFoodz management to its employees and the   2 marks
company then kept to the advertised delivery time of 30 minutes and cut the number of free deliveries to 0.1%. If the original mean of 25 minutes was maintained, calculate the new standard deviation of delivery times.

Question 19 (7 marks) ●●●

The volume of liquid in a 1.25 L bottle of soft drink is normally distributed with a mean of 1259.6 mL and a standard deviation of d mL.

a If $d = 6$, calculate the probability that a bottle will contain less than 1.25 L.   2 marks

b If $d = 15$, the volume of liquid in 90% of the soft drink bottles exceeds x L.   2 marks
Find the value of x, correct to two decimal places.

c The proportion of soft drink bottles in which the volume of liquid is less than 1.2 L is 0.01.   3 marks
Find the value of d, correct to one decimal place.

Question 20 (10 marks) ●●●

A random variable X has the following probability density function:

$$f(x) = \begin{cases} 3a & 0 \le x < 2 \\ a(x-5)(1-x) & 2 \le x \le b \text{ and } 3 < b \le 5 \\ 0 & \text{elsewhere} \end{cases}$$

a Find $P(1 < X < 3)$ in terms of a.   3 marks

b Given $b = 5$, sketch the graph of $y = f(x)$. Clearly state the coordinates of the endpoints and   2 marks
any maximum or minimum points, giving your answers in terms of a.

c Find the value of a.   2 marks

d Show, by integration, that the median of X, m, occurs when $2m^3 - 18m^2 + 30m + 5 = 0$.   3 marks

Practice set 1

Worked solutions

1 B

2 B

$45 - 0.05 = 44.95\,\text{cm}$

$45 + 0.05 = 45.05\,\text{cm}$

1 standard deviation from the mean = 68%

3 A

French

$$z = \frac{x - \mu}{\sigma}$$

$$= \frac{82 - 70}{8}$$

$$= 1$$

Commerce

$$z = \frac{80 - 65}{5}$$

$$= 3$$

Music

$$z = \frac{74 - 50}{12}$$

$$= 2$$

John has the lowest z-score for French and the highest z-score for Commerce.

4 C

z-scores do not directly represent heights.

5 D

$$\frac{1}{6}(a - 2) = 1$$

$$a - 2 = 6$$

$$a = 8$$

6 C

7 B

The mode is the highest point, occurring at $x \approx 7.7$.

8 A

$$z = \frac{x - \mu}{\sigma}$$

$$-2.5 = \frac{x - 71}{4}$$

$$-10 = x - 71$$

$$x = 61$$

9 A

Want $z > 1$:

$$100\% - 50\% - \frac{68\%}{2} = 16\%$$

10 A

63 is 2 standard deviations below the mean, 75 is 1 standard deviation above the mean.

Want $z < -2$ and $z > 1$:

$$z < -2\text{: } \frac{100\% - 95\%}{2} = 2.5\%$$

$z > 1$: 16% from Question 9

$2.5\% + 16\% = 18.5\%$

11 C

$$\int_0^6 ax^2\, dx = \left[\frac{ax^3}{3}\right]_0^6 = 1$$

$$1 = \frac{a \times 6^3}{3} - 0$$

$$= \frac{a \times 216}{3}$$

$$a = \frac{3}{216}$$

$$= \frac{1}{72}$$

12 B

$$P(1.14 \le Z \le 1.40) = P(Z \le 1.40) - P(Z \le 1.14)$$

$$= 0.9192 - 0.8729$$

$$= 0.0463$$

13 C

4 standard deviations between $z = -1$ and $z = 3$

$41 - 23 = 18$, $18 \div 4 = 4.5$

23	27.5	32	36.5	41
$z = -1$	$z = 0$	$z = 1$	$z = 2$	$z = 3$

$\mu = 27.5$

14 A

$$\int_{-3}^{0} p\, dx + \int_{1}^{12p} 3p\, dx = \left[px\right]_{-3}^{0} + \left[3px\right]_{1}^{12p} = 1$$

$$1 = 0 - (-3p) + 3p \times 12p - 3p$$

$$= 3p + 36p^2 - 3p$$

$$= 36p^2$$

$$p = \frac{1}{36}$$

$$= \frac{1}{6}$$

15 B

$$\frac{1}{9}\int_{3.5}^{6} 5x - x^2\, dx = \frac{1}{9}\left[\frac{5x^2}{2} - \frac{x^3}{3}\right]_{3.5}^{6}$$

$$= \frac{1}{9}\left(18 - \frac{6^3}{3} - \frac{5}{2}\times 3.5^2 + \frac{3.5^3}{3}\right)$$

$$= \frac{1}{9}\times\frac{5}{3}$$

$$= \frac{5}{27}$$

16 D

A z-score > 3 has a lower probability than a z-score > 2.

17 C

For the 80th percentile, look for 0.8000 in the table.

When $z = 0.84$, 0.7995.
When $z = 0.85$, 0.8023.

Closest to 0.8 is $z = 0.84$.

$$z = \frac{x - \mu}{\sigma}$$

$$0.84 = \frac{x - 162}{9.5}$$

$$0.84\times 9.5 + 162 = x$$

$$x = 169.98$$

$$\approx 170\,\text{cm}$$

$$\approx 1.7\,\text{m}$$

18 C

$$\int_{26}^{28}\frac{ax}{54}\, dx = 1$$

$$\left[\frac{ax^2}{108}\right]_{26}^{28} = 1$$

$$\frac{a}{108}(28^2 - 26^2) = 1$$

$$\frac{a}{108}\times 108 = 1$$

$$a = 1$$

$$P(26.5 \le x \le 27.5) = \int_{26.5}^{27.5}\frac{x}{54}\, dx$$

$$= \left[\frac{x^2}{108}\right]_{26.5}^{27.5}$$

$$= \frac{1}{108}(27.5^2 - 26.5^2)$$

$$= \frac{54}{108}$$

$$= \frac{1}{2}$$

19 C

Weight of melon is between the 10th and 25th percentile.

$$P(z < -1) = \frac{100\% - 68\%}{2} = 16\% < 25\%$$

$$P(z < -2) = \frac{100\% - 95\%}{2} = 2.5\% < 10\%$$

So it must be graph C.

20 B

$$\int_{k}^{k+1}\cos x + 1\, dx = \left[\sin x + x\right]_{k}^{k+1} = 1$$

$$\sin(k+1) + k + 1 - \sin k - k = 1$$

$$\sin(k+1) - \sin k = 0$$

$$\sin(k+1) - \sin(\pi - k) = 0$$

$$\sin(k+1) = \sin(\pi - k)$$

so $k + 1 = \pi - k$

$$2k = \pi - 1$$

$$k = \frac{\pi - 1}{2}$$

Practice set 2

Worked solutions

Question 1

a i $P(\le 36 \text{ min}) = 0.02 + 0.04 \times 2 = 0.1$

ii $P(35 \le x \le 39 \text{ min})$
$= 0.04 + 0.02 + 0.08 + 0.12 = 0.26$

b $P(\le 38 \text{ min}) = 0.02 \times 2 + 0.04 \times 2 + 0.08 = 0.2$

$P(\text{on time or early on 3 consecutive days})$
$= 0.2^3 = 0.008$

Question 2

$z = \dfrac{151 - 166}{6}$
$= -2.5$

Question 3

Mode = 2 (gives highest point on graph)

Question 4

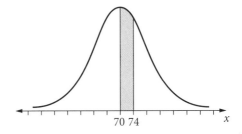

$\mu = 70,\ \sigma = 8$

$z = \dfrac{74 - 70}{8}$

$= \dfrac{1}{2}$

$P\left(z = \dfrac{1}{2}\right) = 0.6915 - 0.5 = 0.1915$

Question 5

a 2 standard deviations from the mean = 95%

Beyond 2 standard deviations from the mean
$= 100 - 95 = 5\%$

b $0.05 \times 1200 = 60$ packets of flour rejected.

Question 6

a

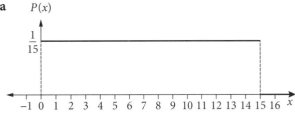

b $\dfrac{7}{15}$ **c** $\dfrac{3}{15} = \dfrac{1}{5}$ **d** $\dfrac{5}{15} = \dfrac{1}{3}$

Question 7

a i

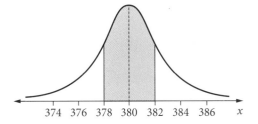

1 standard deviation from the mean = 68%

ii

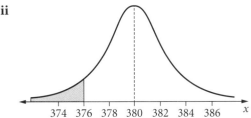

$95 \div 2 = 47.5\%$
$50 - 47.5 = 2.5\%$

iii

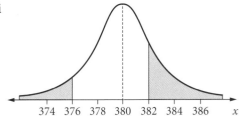

Lower shaded area: (from part **ii**) = 2.5%

Upper shaded area:
$68 \div 2 = 34\%$
$50 - 34 = 16\%$

Total = 2.5 + 16 = 18.5%

b From part **a ii**: 100 − 2.5 = 97.5% of the cans
contain more than 376 mL of soft drink.
This indicates that the manufacturers are
correct, as the cans show a capacity of 375 mL.

Question 8

a $\mu = 35$, $\sigma = 5$

$$z = \frac{30 - 35}{5}$$

$$= -\frac{5}{5}$$

$$= -1$$

$$P(z \geq -1) = 1 - 0.1587 = 0.8413$$

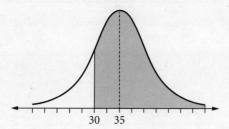

b $\mu = 35$, $\sigma = 5$

$$z = \frac{28 - 35}{5}$$

$$= -\frac{7}{5}$$

$$= -1.4$$

$$P(z \leq -1.4) = 0.0808$$

$$z = \frac{38 - 35}{5}$$

$$= \frac{3}{5}$$

$$= 0.6$$

$$P(z \leq 0.6) = 0.7257$$

$$P(28 < X < 38) = 0.7257 - 0.0808 = 0.6449$$

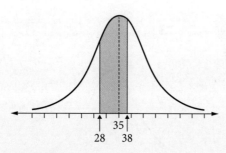

Question 9

Science z-score: $\mu = 75$, $\sigma = 8$

$$z = \frac{81 - 75}{8}$$

$$= 0.75$$

Geography z-score: $\mu = 69$, $\sigma = 6.5$

$$z = \frac{74 - 69}{6.5}$$

$$= 0.769$$

Ashna was not correct. Her Geography mark is better as it has a higher z-score.

Question 10

a

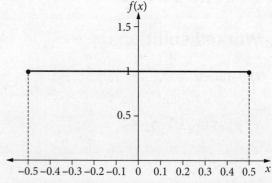

b $P(X > 0.45) = 0.05$
(area under graph from $x = 0.45$ to 0.5)

c $P(X > 0.45) = P(X < -0.45) = 0.05$

$$P(X < -0.45 \mid X < 0) = \frac{P(X < -0.45 \cap X < 0)}{P(X < 0)}$$

$$= \frac{P(X < -0.45)}{P(X < 0)}$$

$$= \frac{0.05}{0.5}$$

$$= 0.1$$

Question 11

a **i** Let X be the weight of tinned fruit.
$P(X < 200) = 0.1$ when $z = -1.28$ (from table)

$$z = \frac{x - \mu}{\sigma}$$

$$= \frac{200 - \mu}{7.8}$$

$$\frac{200 - \mu}{7.8} = -1.28$$

$$\mu = 209.984\ldots$$

$$\approx 210 \text{ grams}$$

ii $z = \dfrac{225 - 210}{7.8}$

$$= 1.923\,07\ldots$$

$$= 1.92$$

$$P(X > 225) = P(Z > 1.92) = 1 - P(Z < 1.92)$$

$$= 1 - 0.9726$$

$$= 0.0274$$

Percentage of cans = $0.0274 \times 100\% = 2.74\%$

b $z = \dfrac{x - \mu}{\sigma}$

$= \dfrac{210 - 205}{\sigma}$

$= \dfrac{5}{\sigma}$

When $-2.33 < z < 2.33$,
98% of cans contain 200–210 g.

$\dfrac{5}{\sigma} = 2.33$

$\dfrac{5}{2.33} = \sigma$

$\sigma = 2.145\,922\ldots$

$\approx 2.15\,\text{g}$

Question 12

a $\displaystyle\int_0^2 ax^2(x - 2)\,dx = 1$

$\displaystyle\int_0^2 ax^3 - 2ax^2\,dx = 1$

$\left[\dfrac{ax^4}{4} - \dfrac{2ax^3}{3}\right]_0^2 = 1$

$\dfrac{16a}{4} - \dfrac{2a}{3} \times 8 = 1$

$4a - \dfrac{16a}{3} = 1$

$\dfrac{12a - 16a}{3} = 1$

$\dfrac{-4a}{3} = 1$

$a = -\dfrac{3}{4}$

b $\displaystyle\int_{12}^2 -\dfrac{3}{4}x^2(x - 2)\,dx$

$= \left[\dfrac{-3x^4}{16} + \dfrac{\cancel{6}^{\cancel{3}}}{\cancel{4}_2} \times \dfrac{x^3}{\cancel{3}}\right]_{1.2}^2$

$= \left[\dfrac{-3x^4}{16} + \dfrac{x^3}{2}\right]_{1.2}^2$

$= \left(\dfrac{-3 \times 2^4}{16} + \dfrac{2^3}{2}\right) - \left(\dfrac{-3 \times 1.2^4}{16} + \dfrac{1.2^3}{2}\right)$

$= -3 + 4 + \dfrac{243}{625} - \dfrac{108}{125}$

$= 0.5248$

c To find the median, check:

$\displaystyle\int_0^{1.2285} -\dfrac{3}{4}x^2(x - 2)\,dx = \dfrac{1}{2}$

$\left[\dfrac{-3x^4}{16} + \dfrac{x^3}{2}\right]_0^{1.2285} = \dfrac{-3 \times 1.2285^4}{16} + \dfrac{1.2285^3}{2} - 0$

$= 0.499\,960\ldots$

$= \dfrac{1}{2}$ as required

Question 13

a 172 is 2 standard deviations above 166.

$\dfrac{100\% - 95\%}{2} = 2.5\%$

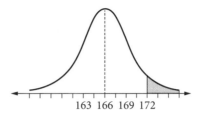

163 166 169 172

b

163 166 169 172

163 is 1 standard deviation below 166.

$\dfrac{68\%}{2} + \dfrac{95\%}{2} = 81.5\%$

$81.5\% \times 215 = 175.225 \approx 175$ students

Question 14

a From equation, when $t = 1, f(1) = \dfrac{1}{3}$.

Area of triangle: base $= 1$, height $= \dfrac{1}{3}$

$A = \dfrac{1}{2} \times 1 \times \dfrac{1}{3} = \dfrac{1}{6}$

$P(t < 1) = \dfrac{1}{6}$

b $P(1 \le t \le 3) = ?$

When $0 < t < 1, A = \dfrac{1}{6}$ from part **a**.

When $t = 3, f(3) = \dfrac{1}{5}$.

When $3 < t < 4, A = \dfrac{1}{2} \times 1 \times \dfrac{1}{5} = \dfrac{1}{10}$.

So, by subtracting areas, the probability
a person waits between 1 and 3 hours is:

$P(1 \le t \le 3) = 1 - \dfrac{1}{6} - \dfrac{1}{10}$

$= \dfrac{11}{15}$

Question 15

a $F(t) = \begin{cases} 0 & t \le 1 \\ \frac{1}{4}(t - 1)^2 & 1 < t \le 2 \\ \frac{1}{16}(7t - t^3 - 8t) & 2 < t \le 4 \\ 1 & t > 4 \end{cases}$

b For the 20th percentile, $1 < t \leq 2$.

$$\frac{1}{4}(t-1)^2 = 0.2$$

$$(t-1)^2 = 0.8$$

$$t - 1 = \pm\sqrt{0.8}$$

$$t = 1 \pm \sqrt{0.8}$$

$$= 1.8944\ldots \text{ or } 0.105\,57\ldots$$

$$= 1.894\,\text{min} \quad (1 < t \leq 2)$$

c $\quad 1 - F(1.5) = 1 - \left(\frac{1}{4} \times 1.5^2 - \frac{1}{2} \times 1.5 + \frac{1}{4}\right)$

$$= \frac{15}{16}$$

Question 16

a $\quad \int_0^k \cos x \, dx = 1$

$$\left[\sin x\right]_0^k = 1$$

$$\sin k - \sin 0 = 1$$

$$\sin k = 1$$

$$k = \frac{\pi}{2}$$

b $\quad \int_0^1 \cos x \, dx = \left[\sin x\right]_0^1$

$$= \sin 1 - \sin 0$$

$$= 0.841\,470\ldots$$

$$\approx 0.8415$$

> **Hint**
> Make sure your calculator is in radian mode.

Question 17

$$\int_1^4 Ax + B \, dx = 1$$

$$\left[\frac{Ax^2}{2} + Bx\right]_1^4 = 1$$

$$\left[\left(\frac{4^2 A}{2} + 4B\right) - \left(\frac{A}{2} + B\right)\right] = 1$$

$$8A + 4B - \frac{A}{2} - B = 1$$

$$\frac{15A}{2} + 3B = 1$$

$$15A + 6B = 2 \quad [1]$$

$$\int_1^2 Ax + B \, dx = \frac{1}{2}$$

$$\left[\frac{Ax^2}{2} + Bx\right]_1^2 = \frac{1}{2}$$

$$\left[\left(\frac{2^2 A}{2} + 2B\right) - \left(\frac{A}{2} + B\right)\right] = \frac{1}{2}$$

$$2A + 2B - \frac{A}{2} - B = \frac{1}{2}$$

$$\frac{3A}{2} + B = \frac{1}{2}$$

$$3A + 2B = 1 \quad [2]$$

$3 \times [2]:\qquad 9A + 6B = 3 \qquad [3]$

$[1] - [3]:\qquad 6A = -1$

$$A = -\frac{1}{6}$$

Substitute into [2]:

$$3\left(-\frac{1}{6}\right) + 2B = 1$$

$$-\frac{1}{2} + 2B = 1$$

$$2B = \frac{3}{2}$$

$$B = \frac{3}{4}$$

Question 18

a $\quad z = \dfrac{30 - 25}{4}$

$$= 1.25$$

Reading from the normal distribution table, $P(z \leq 1.25) = 0.8944$.

$$P(T > 30) = P(z > 1.25) = 1 - 0.8944 = 0.1056$$

b $0.1\% = 0.001$, when $z \geq 3.10$.

$$z = \frac{x - \mu}{\sigma}$$

$$3.1 = \frac{x - 25}{4}$$

$$x = 25 + 3.1 \times 4$$

$$= 37.4 \text{ minutes}$$

The new delivery time would be 37.4 minutes.

c $\quad z = \dfrac{x - \mu}{\sigma}$

$$3.10 = \frac{30 - 25}{\sigma}$$

$$\sigma = \frac{30 - 25}{3.10}$$

$$= 1.6129\ldots$$

$$\approx 1.6$$

The new standard deviation would be 1.6 minutes.

WORKED SOLUTIONS

Question 19

a $z = \dfrac{1250 - 1259.6}{6}$

$= \dfrac{-9.6}{6}$

$= -1.6$

$P(z < -1.6) = 0.0548$

b Reading from normal distribution table, $z = -1.28$ for 90% of the soft drink bottles to exceed x L.

$z = \dfrac{x - \mu}{\sigma}$

$-1.28 = \dfrac{x - 1259.6}{15}$

$-1.28 \times 15 = x - 1259.6$

$x = 1259.6 - 19.2$

$= 1240.2 \text{ mL}$

$= 1.24 \text{ L}$

c $z = \dfrac{x - \mu}{\sigma}$

$= \dfrac{1200 - 1259.6}{d}$

$= \dfrac{-59.6}{d}$

Reading from normal distribution table, $z = -2.33$ for proportion of bottles with less than 1.2 L of liquid.

$-2.33 = \dfrac{-59.6}{d}$

$d = \dfrac{-59.6}{-2.33}$

$= 25.5793\ldots$

$= 25.6$

Question 20

a $P(1 < X < 3) = \displaystyle\int_1^2 3a\,dx + \int_2^3 a(6x - x^2 - 5)\,dx$

$= \left[3ax\right]_1^2 + a\left[3x^2 - \dfrac{x^3}{3} - 5x\right]_2^3$

$= 3a \times 2 - 3a + a\left[27 - 9 - 15 - \left(12 - \dfrac{8}{3} - 10\right)\right]$

$= 3a + a\left(\dfrac{11}{3}\right)$

$= \dfrac{20a}{3}$

b

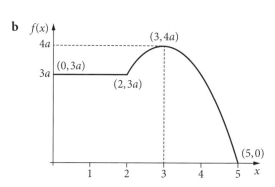

c $p(0 \le X \le 5) = \displaystyle\int_0^2 3a\,dx + \int_2^5 a(6x - x^2 - 5)\,dx$

$= \left[3ax\right]_0^2 + a\left[3x^2 - \dfrac{x^3}{3} - 5x\right]_2^5$

$= 6a + a\left[75 - \dfrac{125}{3} - 25 - \left(12 - \dfrac{8}{3} - 10\right)\right]$

$= 6a + 9a$

$= 15a$

$15a = 1$

$a = \dfrac{1}{15}$

d
$$\int_0^m f(x)\,dx = 0.5$$

$$6 \times \frac{1}{15} + \frac{1}{15}\left[\frac{-x^3}{3} + 3x^2 - 5x\right]_2^m = \frac{1}{2}$$

$$\frac{2}{5} + \frac{1}{15}\left[\frac{-m^3}{3} + 3m^2 - 5m - \left(-\frac{8}{3} + 12 - 10\right)\right] = \frac{1}{2}$$

$$\frac{-m^3}{3} + 3m^2 - 5m + \frac{2}{3} = \frac{3}{2}$$

$$-2m^3 + 18m^2 - 30m - 4 = 9$$

$$-2m^3 + 18m^2 - 30m - 5 = 0$$

$$2m^3 - 18m^2 + 30m + 5 = 0 \quad \text{as required}$$

HSC exam topic grid (2011–2020)

This table shows the coverage of this topic in past HSC exams by question number. The past exams can be downloaded from the NESA website (www.educationstandards.nsw.edu.au) by selecting 'Year 11 – Year 12', 'HSC exam papers'. NESA marking feedback and guidelines can also be found there.

This topic was introduced to the Mathematics Advanced course in 2020. The normal distribution and z-scores are common content with Mathematics Standard 2. In the table below, * refers to past HSC questions (2011–2019) in Mathematics Standard. Before 2019, 'Mathematics Standard 2' was called 'Mathematics General 2', and before 2014, 'General Mathematics'. For these exams, select 'Year 11 – Year 12', 'Resources archive', 'HSC exam papers archive'.

	Discrete random variables (Year 11)	Continuous random variables	The normal distribution	z-scores
2011*				27(c)
2012*			29(b)	
2013*			20	29(b)
2014*			24	
2015*			20	**28(b)**
2016*			13	
2017*			29(d)	13
2018*			23	27(e)
2019*			15	38
2020 new course		**23**	**9, 28**	**3**

Questions in **bold** can be found in this chapter.

HSC exam reference sheet

Mathematics Advanced, Extension 1 and Extension 2

© NSW Education Standards Authority

Note: Unlike the actual HSC reference sheet, this sheet indicates which formulas are Mathematics Extension 1 and 2.

Measurement

Length

$$l = \frac{\theta}{360} \times 2\pi r$$

Area

$$A = \frac{\theta}{360} \times \pi r^2$$

$$A = \frac{h}{2}(a + b)$$

Surface area

$$A = 2\pi r^2 + 2\pi rh$$

$$A = 4\pi r^2$$

Volume

$$V = \frac{1}{3}Ah$$

$$V = \frac{4}{3}\pi r^3$$

Functions

$$x = \frac{-b \pm \sqrt{b^2 - 4ac}}{2a}$$

For $ax^3 + bx^2 + cx + d = 0$:*   *EXT1

$$\alpha + \beta + \gamma = -\frac{b}{a}$$

$$\alpha\beta + a\gamma + \beta\gamma = \frac{c}{a}$$

$$\text{and } \alpha\beta\gamma = -\frac{d}{a}$$

Relations

$$(x - h)^2 + (y - k)^2 = r^2$$

Financial Mathematics

$$A = P(1 + r)^n$$

Sequences and series

$$T_n = a + (n - 1)d$$

$$S_n = \frac{n}{2}[2a + (n - 1)d] = \frac{n}{2}(a + l)$$

$$T_n = ar^{n-1}$$

$$S_n = \frac{a(1 - r^n)}{1 - r} = \frac{a(r^n - 1)}{r - 1}, r \neq 1$$

$$S = \frac{a}{1 - r}, |r| < 1$$

Logarithmic and Exponential Functions

$$\log_a a^x = x = a^{\log_a x}$$

$$\log_a x = \frac{\log_b x}{\log_b a}$$

$$a^x = e^{x \ln a}$$

Trigonometric Functions

$$\sin A = \frac{\text{opp}}{\text{hyp}}, \quad \cos A = \frac{\text{adj}}{\text{hyp}}, \quad \tan A = \frac{\text{opp}}{\text{adj}}$$

$$A = \frac{1}{2}ab\sin C$$

$$\frac{a}{\sin A} = \frac{b}{\sin B} = \frac{c}{\sin C}$$

$$c^2 = a^2 + b^2 - 2ab\cos C$$

$$\cos C = \frac{a^2 + b^2 - c^2}{2ab}$$

$$l = r\theta$$

$$A = \frac{1}{2}r^2\theta$$

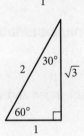

Trigonometric identities

$$\sec A = \frac{1}{\cos A}, \quad \cos A \neq 0$$

$$\operatorname{cosec} A = \frac{1}{\sin A}, \quad \sin A \neq 0$$

$$\cot A = \frac{\cos A}{\sin A}, \quad \sin A \neq 0$$

$$\cos^2 x + \sin^2 x = 1$$

Compound angles*

$$\sin(A + B) = \sin A\cos B + \cos A\sin B$$

$$\cos(A + B) = \cos A\cos B - \sin A\sin B$$

$$\tan(A + B) = \frac{\tan A + \tan B}{1 - \tan A\tan B}$$

If $t = \tan\dfrac{A}{2}$, then $\sin A = \dfrac{2t}{1 + t^2}$

$$\cos A = \frac{1 - t^2}{1 + t^2}$$

$$\tan A = \frac{2t}{1 - t^2}$$

$$\cos A\cos B = \frac{1}{2}\big[\cos(A - B) + \cos(A + B)\big]$$

$$\sin A\sin B = \frac{1}{2}\big[\cos(A - B) - \cos(A + B)\big]$$

$$\sin A\cos B = \frac{1}{2}\big[\sin(A + B) + \sin(A - B)\big]$$

$$\cos A\sin B = \frac{1}{2}\big[\sin(A + B) - \sin(A - B)\big]$$

$$\sin^2 nx = \frac{1}{2}(1 - \cos 2nx)$$

$$\cos^2 nx = \frac{1}{2}(1 + \cos 2nx)$$

*EXT1

Statistical Analysis

$$z = \frac{x - \mu}{\sigma}$$

An outlier is a score less than $Q_1 - 1.5 \times \text{IQR}$ or more than $Q_3 + 1.5 \times \text{IQR}$

Normal distribution

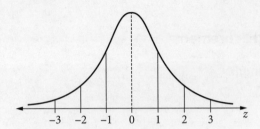

- approximately 68% of scores have z-scores between -1 and 1
- approximately 95% of scores have z-scores between -2 and 2
- approximately 99.7% of scores have z-scores between -3 and 3

Discrete random variables

$$E(X) = \mu$$

$$\operatorname{Var}(X) = E\big[(X - \mu)^2\big] = E(X^2) - \mu^2$$

Probability

$$P(A \cap B) = P(A)P(B)$$

$$P(A \cup B) = P(A) + P(B) - P(A \cap B)$$

$$P(A|B) = \frac{P(A \cap B)}{P(B)}, \quad P(B) \neq 0$$

Continuous random variables

$$P(X \leq r) = \int_a^r f(x)\,dx$$

$$P(a < X < b) = \int_a^b f(x)\,dx$$

Binomial distribution*

$$P(X = r) = {}^nC_r\, p^r (1 - p)^{n-r}$$

$$X \sim \text{Bin}(n, p)$$
$$\Rightarrow P(X = x)$$
$$= \binom{n}{x} p^x (1 - p)^{n-x}, \quad x = 0, 1, \ldots, n$$

$$E(X) = np$$

$$\operatorname{Var}(X) = np(1 - p)$$

9780170459228

Differential Calculus

Function	Derivative
$y = f(x)^n$	$\dfrac{dy}{dx} = nf'(x)\left[f(x)\right]^{n-1}$
$y = uv$	$\dfrac{dy}{dx} = u\dfrac{dv}{dx} + v\dfrac{du}{dx}$
$y = g(u)$ where $u = f(x)$	$\dfrac{dy}{dx} = \dfrac{dy}{du} \times \dfrac{du}{dx}$
$y = \dfrac{u}{v}$	$\dfrac{dy}{dx} = \dfrac{v\dfrac{du}{dx} - u\dfrac{dv}{dx}}{v^2}$
$y = \sin f(x)$	$\dfrac{dy}{dx} = f'(x)\cos f(x)$
$y = \cos f(x)$	$\dfrac{dy}{dx} = -f'(x)\sin f(x)$
$y = \tan f(x)$	$\dfrac{dy}{dx} = f'(x)\sec^2 f(x)$
$y = e^{f(x)}$	$\dfrac{dy}{dx} = f'(x)e^{f(x)}$
$y = \ln f(x)$	$\dfrac{dy}{dx} = \dfrac{f'(x)}{f(x)}$
$y = a^{f(x)}$	$\dfrac{dy}{dx} = (\ln a)f'(x)a^{f(x)}$
$y = \log_a f(x)$	$\dfrac{dy}{dx} = \dfrac{f'(x)}{(\ln a)f(x)}$
$y = \sin^{-1} f(x)$	$\dfrac{dy}{dx} = \dfrac{f'(x)}{\sqrt{1 - \left[f(x)\right]^2}}$ *
$y = \cos^{-1} f(x)$	$\dfrac{dy}{dx} = -\dfrac{f'(x)}{\sqrt{1 - \left[f(x)\right]^2}}$ *
$y = \tan^{-1} f(x)$	$\dfrac{dy}{dx} = \dfrac{f'(x)}{1 + \left[f(x)\right]^2}$ *

Integral Calculus

$$\int f'(x)\left[f(x)\right]^n dx = \frac{1}{n+1}\left[f(x)\right]^{n+1} + c$$
$$\text{where } n \neq -1$$

$$\int f'(x)\sin f(x)\, dx = -\cos f(x) + c$$

$$\int f'(x)\cos f(x)\, dx = \sin f(x) + c$$

$$\int f'(x)\sec^2 f(x)\, dx = \tan f(x) + c$$

$$\int f'(x)e^{f(x)}\, dx = e^{f(x)} + c$$

$$\int \frac{f'(x)}{f(x)}\, dx = \ln|f(x)| + c$$

$$\int f'(x)a^{f(x)}\, dx = \frac{a^{f(x)}}{\ln a} + c$$

$$\int \frac{f'(x)}{\sqrt{a^2 - \left[f(x)\right]^2}}\, dx = \sin^{-1}\frac{f(x)}{a} + c \ *$$

$$\int \frac{f'(x)}{a^2 + \left[f(x)\right]^2}\, dx = \frac{1}{a}\tan^{-1}\frac{f(x)}{a} + c \ *$$

$$\int u\frac{dv}{dx}dx = uv - \int v\frac{du}{dx}\, dx \ **$$

$$\int_a^b f(x)\, dx$$
$$\approx \frac{b-a}{2n}\left\{f(a) + f(b) + 2\left[f(x_1) + \cdots + f(x_{n-1})\right]\right\}$$
where $a = x_0$ and $b = x_n$

*EXT1, **EXT2

Combinatorics*

$$^nP_r = \frac{n!}{(n-r)!}$$

$$\binom{n}{r} = {}^nC_r = \frac{n!}{r!(n-r)!}$$

$$(x+a)^n = x^n + \binom{n}{1}x^{n-1}a + \cdots + \binom{n}{r}x^{n-r}a^r + \cdots + a^n$$

Vectors*

$$\left|\underset{\sim}{u}\right| = \left|x\underset{\sim}{i} + y\underset{\sim}{j}\right| = \sqrt{x^2 + y^2}$$

$$\underset{\sim}{u} \cdot \underset{\sim}{v} = \left|\underset{\sim}{u}\right|\left|\underset{\sim}{v}\right|\cos\theta = x_1 x_2 + y_1 y_2,$$
where $\underset{\sim}{u} = x_1\underset{\sim}{i} + y_1\underset{\sim}{j}$
and $\underset{\sim}{v} = x_2\underset{\sim}{i} + y_2\underset{\sim}{j}$

$$\underset{\sim}{r} = \underset{\sim}{a} + \lambda\underset{\sim}{b}**$$

Complex Numbers**

$$z = a + ib = r(\cos\theta + i\sin\theta)$$
$$= re^{i\theta}$$

$$\left[r(\cos\theta + i\sin\theta)\right]^n = r^n(\cos n\theta + i\sin n\theta)$$
$$= r^n e^{in\theta}$$

Mechanics**

$$\frac{d^2x}{dt^2} = \frac{dv}{dt} = v\frac{dv}{dx} = \frac{d}{dx}\left(\frac{1}{2}v^2\right)$$

$$x = a\cos(nt + \alpha) + c$$

$$x = a\sin(nt + \alpha) + c$$

$$\ddot{x} = -n^2(x - c)$$

*EXT1, **EXT2

9780170459228

Index